Selected Titles in This Series

230 **Ezra Getzler and Mikhail Kapranov, Editors,** Higher category theory, 1998

229 **Edward L. Green and Birge Huisgen-Zimmermann, Editors,** Trends in the representation theory of finite dimensional algebras, 1998

228 **Liming Ge, Huaxin Lin, Zhong-Jin Ruan, Dianzhou Zhang, and Shuang Zhang, Editors,** Operator algebras and operator theory, 1999

227 **John McCleary, Editor,** Higher homotopy structures in topology and mathematical physics, 1999

226 **Luis A. Caffarelli and Mario Milman, Editors,** Monge Ampère equation: Applications to geometry and optimization, 1999

225 **Ronald C. Mullin and Gary L. Mullen, Editors,** Finite fields: Theory, applications, and algorithms, 1999

224 **Sang Geun Hahn, Hyo Chul Myung, and Efim Zelmanov, Editors,** Recent progress in algebra, 1999

223 **Bernard Chazelle, Jacob E. Goodman, and Richard Pollack, Editors,** Advances in discrete and computational geometry, 1999

222 **Kang-Tae Kim and Steven G. Krantz, Editors,** Complex geometric analysis in Pohang, 1999

221 **J. Robert Dorroh, Giséle Ruiz Goldstein, Jerome A. Goldstein, and Michael Mudi Tom, Editors,** Applied analysis, 1999

220 **Mark Mahowald and Stewart Priddy, Editors,** Homotopy theory via algebraic geometry and group representations, 1998

219 **Marc Henneaux, Joseph Krasil′shchik, and Alexandre Vinogradov, Editors,** Secondary calculus and cohomological physics, 1998

218 **Jan Mandel, Charbel Farhat, and Xiao-Chuan Cai, Editors,** Domain decomposition methods 10, 1998

217 **Eric Carlen, Evans M. Harrell, and Michael Loss, Editors,** Advances in differential equations and mathematical physics, 1998

216 **Akram Aldroubi and EnBing Lin, Editors,** Wavelets, multiwavelets, and their applications, 1998

215 **M. G. Nerurkar, D. P. Dokken, and D. B. Ellis, Editors,** Topological dynamics and applications, 1998

214 **Lewis A. Coburn and Marc A. Rieffel, Editors,** Perspectives on quantization, 1998

213 **Farhad Jafari, Barbara D. MacCluer, Carl C. Cowen, and A. Duane Porter, Editors,** Studies on composition operators, 1998

212 **E. Ramírez de Arellano, N. Salinas, M. V. Shapiro, and N. L. Vasilevski, Editors,** Operator theory for complex and hypercomplex analysis, 1998

211 **Józef Dodziuk and Linda Keen, Editors,** Lipa's legacy: Proceedings from the Bers Colloquium, 1997

210 **V. Kumar Murty and Michel Waldschmidt, Editors,** Number theory, 1998

209 **Steven Cox and Irena Lasiecka, Editors,** Optimization methods in partial differential equations, 1997

208 **Michel L. Lapidus, Lawrence H. Harper, and Adolfo J. Rumbos, Editors,** Harmonic analysis and nonlinear differential equations: A volume in honor of Victor L. Shapiro, 1997

207 **Yujiro Kawamata and Vyacheslav V. Shokurov, Editors,** Birational algebraic geometry: A conference on algebraic geometry in memory of Wei-Liang Chow (1911–1995), 1997

206 **Adam Korányi, Editor,** Harmonic functions on trees and buildings, 1997

(See the AMS catalog for earlier titles)

Higher Category Theory

CONTEMPORARY
MATHEMATICS

230

Higher Category Theory

Workshop on Higher Category Theory and Physics
March 28–30, 1997
Northwestern University
Evanston, IL

Ezra Getzler
Mikhail Kapranov
Editors

American Mathematical Society
Providence, Rhode Island

This volume is the record of a Workshop on Higher Category Theory and Physics, which took place at Northwestern University, Evanston, IL, on March 28–30, 1997.

1991 *Mathematics Subject Classification.* Primary 18-06, 18D05, 18G50.

Library of Congress Cataloging-in-Publication Data

Conference on Category Theory (1997 : Northwestern University)
High category theory : proceedings of a Conference on Category Theory, March 1997, Northwestern University, Evanston, IL / Ezra Getzler, Mikhail Kapranov, editors.
p. cm. — (Contemporary mathematics, ISSN 0271-4132 ; 230)
Includes bibliographical references.
ISBN 0-8218-1056-1 (alk. paper)
1. Categories (Mathematics)—Congresses. I. Getzler, Ezra, 1962– . II. Kapranov, M. M. (Mikhail M.), 1962– . III. Title. IV. Series: Contemporary mathematics (American Mathematical Society) ; v. 230.
QA169.C64 1997
511.3—dc21 98-32266
CIP

∞ The paper used in this book is acid-free and falls within the guidelines
established to ensure permanence and durability.
Visit the AMS home page at URL: http://www.ams.org/

10 9 8 7 6 5 4 3 2 1 03 02 01 00 99 98

Contents

Preface

The concept of categories, introduced by Eilenberg and Mac Lane in the 1940's, has replaced set theory as the foundation in a large part of mathematics. One reason is that more information is retained if one considers objects in categories rather than their underlying sets. To put it more concisely: look at a set through a magnifying glass and you see a category. The passage from sets to categories involves a change in point of view: in most cases, it is neither meaningful nor necessary to speak about two objects of a category being equal[1]; much better to say that they are isomorphic, and specify the isomorphism.

Soon after the introduction of categories, it was realized that the Hom-sets in a category are frequently in their turn categories, and so on. Iterating n times, one obtains the theory of n-categories. A 0-category is just a set, a 1-category is the same as an ordinary category and an example of a 2-category is provided by the "category" of all 1-categories. However, it turned out to be very difficult to give an "invariant" definition of an n-category (i.e. a definition which at no stage involves the concept of equality of two objects of an ordinary category). This problem was explicitly stated by Grothendieck in his unpublished memoir "Pursuing stacks" (1986). A naive attempt to write down a definition stage by stage quickly leads to seemingly unsurmountable combinatorial difficulties.

In recent years, several definition of (weak) n-categories have been proposed [**1,2,3,4**]. One reason for increased interest in the subject was the discovery of applications in low-dimensional topology and mathematical physics, where monoidal categories (2-categories with one object) have been used for over a decade, and where the recent work has led to the consideration of higher categorical structures.

This volume is the record of the workshop on higher category theory and mathematical physics held at Northwestern University in March 1997 with the aim of making some of these exciting new developments in category theory better known outside the community of experts, as they deserve to be. We especially tried to solicit presentations in the style "Higher categories for the working mathematician." We hope that the present collection of papers will be useful as an introduction to a subject which is intrinsically difficult but holds great promise for the future development of mathematics.

Ezra Getzler, Misha Kapranov
October 1998, Evanston

[1]Compare to these maxims of Wittgenstein: "To say of two things that they are identical is nonsense and to say that one thing is identical with itself is to say nothing at all (...) The identity-sign, therefore, is not an essential constituent of conceptual notation" ([**5**], §5.53).

Bibliography

[1] J. C. Baez and J. Dolan, *Higher-dimensional algebra III: n-categories and the algebra of opetopes,* Adv. Math. **135** (1998), 145-206.

[2] M. A. Batanin, *Monoidal globular categories as a natural environment for the theory of weak n-categories,* Adv. Math. **136** (1998), 39-103.

[3] C. Hermida, M. Makkai and J. Power, *On weak higher-dimensional categories,* preprint, 1997.

[4] Z. Tamsamani, *Sur les notions de n-categorie et de n-groupoïde non-strictes via des ensembles multi-simpliciaux,* preprint `alg-geom/9512006`.

[5] L. Wittgenstein, *Tractatus logico-philosophicus.*

Contemporary Mathematics
Volume **230**, 1998

Categorification

John C. Baez and James Dolan

ABSTRACT. Categorification is the process of finding category-theoretic analogs of set-theoretic concepts by replacing sets with categories, functions with functors, and equations between functions by natural isomorphisms between functors, which in turn should satisfy certain equations of their own, called 'coherence laws'. Iterating this process requires a theory of 'n-categories', algebraic structures having objects, morphisms between objects, 2-morphisms between morphisms and so on up to n-morphisms. After a brief introduction to n-categories and their relation to homotopy theory, we discuss algebraic structures that can be seen as iterated categorifications of the natural numbers and integers. These include tangle n-categories, cobordism n-categories, and the homotopy n-types of the loop spaces $\Omega^k S^k$. We conclude by describing a definition of weak n-categories based on the theory of operads.

1. Introduction

Though the concept it refers to has been lurking in the collective subconscious of mathematics for over a century, gradually becoming conscious, the term 'categorification' was invented only rather recently, by Crane [**19, 20**]. Categorification is the process of finding category-theoretic analogs of ideas phrased in the language of set theory, using the following analogy between set theory and category theory:

elements	objects
equations between elements	isomorphisms between objects
sets	categories
functions	functors
equations between functions	natural isomorphisms between functors

1. Analogy between set theory and category theory

Just as sets have elements, categories have objects. Just as there are functions between sets, there are functors between categories. Interestingly, the proper analog

1991 *Mathematics Subject Classification.* Primary 18G30, 55P15, 55P35, 55P42; Secondary 20L15, 18D05.

of an equation between elements is not an equation between objects, but an isomorphism. More generally, the analog of an equation between functions is a natural isomorphism between functors.

For example, the category FinSet, whose objects are finite sets and whose morphisms are functions, is a categorification of the set $\mathbb{N}$ of natural numbers. The disjoint union and Cartesian product of finite sets correspond to the sum and product in $\mathbb{N}$, respectively. Note that while addition and multiplication in $\mathbb{N}$ satisfy various equational laws such as commutativity, associativity and distributivity, disjoint union and Cartesian product satisfy such laws *only up to natural isomorphism.*

If one studies categorification one soon discovers an amazing fact: many deep-sounding results in mathematics are just categorifications of facts we learned in high school! There is a good reason for this. All along, we have been unwittingly 'decategorifying' mathematics by pretending that categories are just sets. We 'decategorify' a category by forgetting about the morphisms and pretending that isomorphic objects are equal. We are left with a mere set: the set of isomorphism classes of objects.

To understand this, the following parable may be useful. Long ago, when shepherds wanted to see if two herds of sheep were isomorphic, they would look for an explicit isomorphism. In other words, they would line up both herds and try to match each sheep in one herd with a sheep in the other. But one day, along came a shepherd who invented decategorification. She realized one could take each herd and 'count' it, setting up an isomorphism between it and some set of 'numbers', which were nonsense words like 'one, two, three, ... ' specially designed for this purpose. By comparing the resulting numbers, she could show that two herds were isomorphic without explicitly establishing an isomorphism! In short, by decategorifying the category of finite sets, the set of natural numbers was invented.

According to this parable, decategorification started out as a stroke of mathematical genius. Only later did it become a matter of dumb habit, which we are now struggling to overcome by means of categorification. While the historical reality is far more complicated, categorification really has led to tremendous progress in mathematics during the 20th century. For example, Noether revolutionized algebraic topology by emphasizing the importance of homology groups. Previous work had focused on Betti numbers, which are just the dimensions of the rational homology groups. As with taking the cardinality of a set, taking the dimension of a vector space is a process of decategorification, since two vector spaces are isomorphic if and only if they have the same dimension. Noether noted that if we work with homology groups rather than Betti numbers, we can solve more problems, because we obtain invariants not only of spaces, but also of maps. In modern parlance, the nth rational homology is a *functor* defined on the *category* of topological spaces, while the nth Betti number is a mere *function* defined on the *set* of isomorphism classes of topological spaces. Of course, this way of stating Noether's insight is anachronistic, since it came before category theory. Indeed, it was in Eilenberg and Mac Lane's subsequent work on homology that category theory was born!

Decategorification is a straightforward process which typically destroys information about the situation at hand. Categorification, being an attempt to recover this lost information, is inevitably fraught with difficulties. One reason is that when categorifying, one does not merely replace equations by isomorphisms. One also demands that these isomorphisms satisfy some new equations of their own, called 'coherence laws'. Finding the right coherence laws for a given situation is perhaps the trickiest aspect of categorification.

For example, a monoid is a set with a product satisfying the associative law and a unit element satisfying the left and right unit laws. The categorified version of a monoid is a 'monoidal category'. This is a category C with a product $\otimes\colon C \times C \to C$ and unit object $1 \in C$. If we naively impose associativity and the left and right unit laws as equational laws, we obtain the definition of a 'strict' monoidal category. However, the philosophy of categorification suggests instead that we impose them only up to natural isomorphism. Thus, as part of the structure of a 'weak' monoidal category, we specify a natural isomorphism

$$a_{x,y,z}\colon (x \otimes y) \otimes z \to x \otimes (y \otimes z)$$

called the 'associator', together with natural isomorphisms

$$l_x\colon 1 \otimes x \to x,$$

$$r_x\colon x \otimes 1 \to x.$$

Using the associator one can construct isomorphisms between any two parenthesized versions of the tensor product of several objects. However, we really want a *unique* isomorphism. For example, there are 5 ways to parenthesize the tensor product of 4 objects, which are related by the associator as follows:

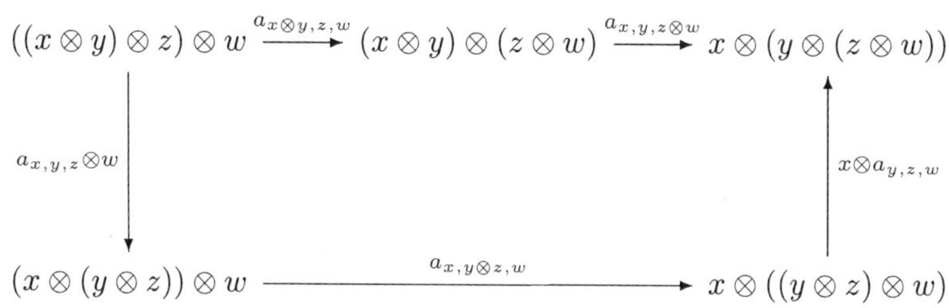

In the definition of a weak monoidal category we impose a coherence law, called the 'pentagon identity', saying that this diagram commutes. Similarly, we impose a coherence law saying that the following diagram built using a, l and r commutes:

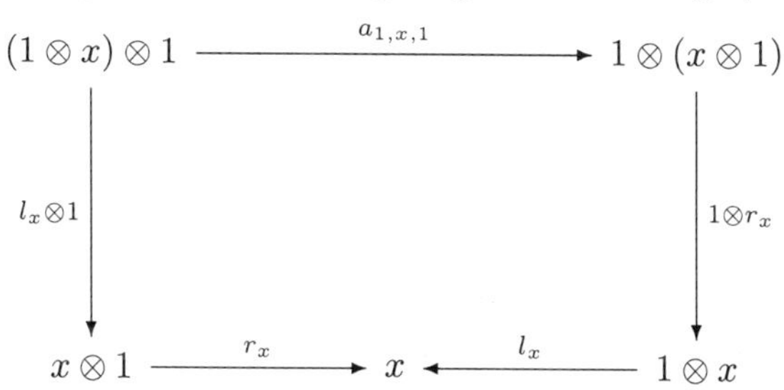

This definition raises an obvious question: how do we know we have found all the right coherence laws? Indeed, what does 'right' even *mean* in this context? Mac Lane's coherence theorem [**45**] gives one answer to this question: the above coherence laws imply that any two isomorphisms built using a, l and r and having the same source and target must be equal.

Further work along these lines allow us to make more precise the sense in which $\mathbb{N}$ is a decategorification of FinSet. For example, just as $\mathbb{N}$ forms a monoid under either addition or multiplication, FinSet becomes a monoidal category under either disjoint union or Cartesian product if we choose the isomorphisms a, l, and r sensibly. In fact, just as $\mathbb{N}$ is a 'rig', satisfying all the ring axioms except those involving additive inverses, FinSet is what one might call a 'rig category'. In other words, it satisfies the rig axioms up to natural isomorphisms satisfying the coherence laws discovered by Kelly [**41**] and Laplaza [**44**], who proved a coherence theorem in this context. Just as the decategorification of a monoidal category is a monoid, the decategorification of any rig category is a rig. In particular, decategorifying the rig category FinSet gives the rig $\mathbb{N}$. This idea is especially important in combinatorics, where the best proof of an identity involving natural numbers is often a 'bijective proof': one that actually establishes an isomorphism between finite sets [**37, 54**].

While coherence laws can sometimes be justified retrospectively by coherence theorems, certain puzzles point to the need for a deeper understanding of the *origin* of coherence laws. For example, suppose we want to categorify the notion of 'commutative monoid'. The strictest possible approach, where we take a strict monoidal category and impose an equational law of the form $x \otimes y = y \otimes x$, is almost completely uninteresting. It is much better to start with a weak monoidal category equipped with a natural isomorphism $B_{x,y}\colon x \otimes y \to y \otimes x$ called the 'braiding', and then impose coherence laws called 'hexagon identities' saying that the following two diagrams commute:

$$\begin{array}{ccccc}
x \otimes (y \otimes z) & \xrightarrow{a^{-1}_{x,y,z}} & (x \otimes y) \otimes z & \xrightarrow{B_{x,y} \otimes z} & (y \otimes x) \otimes z \\
\Big\downarrow {\scriptstyle B_{x,y \otimes z}} & & & & \Big\downarrow {\scriptstyle a_{y,x,z}} \\
(y \otimes z) \otimes x & \xrightarrow[a_{y,z,x}]{} & y \otimes (z \otimes x) & \xleftarrow[y \otimes B_{x,z}]{} & y \otimes (x \otimes z)
\end{array}$$

$$\begin{array}{ccccc}
(x \otimes y) \otimes z & \xrightarrow{a_{x,y,z}} & x \otimes (y \otimes z) & \xrightarrow{x \otimes B_{y,z}} & x \otimes (z \otimes y) \\
\Big\downarrow {\scriptstyle B_{x \otimes y,z}} & & & & \Big\downarrow {\scriptstyle a^{-1}_{x,z,y}} \\
z \otimes (x \otimes y) & \xrightarrow[a^{-1}_{z,x,y}]{} & (z \otimes x) \otimes y & \xleftarrow[B_{x,z} \otimes y]{} & (x \otimes z) \otimes y
\end{array}$$

This gives the definition of a weak 'braided monoidal category'. If we impose an additional coherence law of the form $B^{-1}_{x,y} = B_{y,x}$, we obtain the definition of a 'symmetric monoidal category'. Both of these concepts are very important; which one is 'right' depends on the context. However, neither implies that every pair of parallel morphisms built using the braiding are equal. A good theory of coherence laws must naturally account for these facts.

The deepest insights into such puzzles have traditionally come from topology. In homotopy theory it causes problems to work with spaces equipped with algebraic structures satisfying equational laws, because one cannot transport such structures along homotopy equivalences. It is better to impose laws *only up to homotopy*, with these homotopies satisfying certain coherence laws, but again only up to homotopy, with these higher homotopies satisfying their own higher coherence laws, and so on. Coherence laws thus arise naturally in infinite sequences. For example, Stasheff [**59**] discovered the pentagon identity and a sequence of higher coherence laws for associativity when studying the algebraic structure possessed by a space that is homotopy equivalent to a loop space. Similarly, the hexagon identities arise as part of a sequence of coherence laws for spaces homotopy equivalent to double loop spaces, while the extra coherence law for symmetric monoidal categories arises as part of a sequence for spaces homotopy equivalent to triple loop spaces. The higher coherence laws in these sequences turn out to be crucial when we try to *iterate* the process of categorification.

Starting in the late 1960's, Boardman, Vogt [**13, 14**] and others developed the study of these higher coherence laws into a full-fledged theory of 'homotopy-invariant algebraic structures'. However, we have yet to attain a general and systematic theory of categorification, particularly when it comes to iterated categorification, which requires a good theory of 'n-categories'. The main goal of this paper is to outline our current understanding of categorification and to point out some directions for further study.

The plan of the paper is as follows. In Section 2 we give a quick tour of n-category theory. This theory is just beginning to be developed, and there are various alternative approaches which have not yet been reconciled, but here we leave out most of the technical details and sketch what we expect from any reasonable approach. In Section 3 we discuss in more detail the lessons homotopy theory has for n-category theory. We hope this section can be followed even by those who are not already experts on homotopy theory. In Section 4 we describe some algebraic structures that amount to iterated categorifications of the natural numbers and the integers. A large amount of interesting mathematics emerges from the study of these structures. In Section 5 we summarize our own approach to n-categories.

In a previous paper [**4**] we sketched a program of using n-categories to clarify the relationships between topological quantum field theory and more traditional approaches to algebraic topology. The present paper covers some aspects of this program in more detail, taking advantage of work that has been done in the meantime. Various other aspects are treated in a series of papers entitled 'Higher-Dimensional Algebra' [**2, 5, 6, 8**].

2. n-Categories

One philosophical reason for categorification is that it refines our concept of 'sameness' by allowing us to distinguish between isomorphism and equality. In a set, two elements are either the same or different. In a category, two objects can be 'the same in a way' while still being different. In other words, they can be isomorphic but not equal. Even more importantly, two objects can be the same in more than one way, since there can be different isomorphisms between them. This gives rise to the notion of the 'symmetry group' of an object: its group of automorphisms.

Consider, for example, the fundamental groupoid $\Pi_1(X)$ of a topological space X: the category with points of X as objects and homotopy classes of paths with fixed endpoints as morphisms. This category captures all the homotopy-theoretic information about X in dimensions ≤ 1 — or more precisely, its homotopy 1-type [**11**]. The group of automorphisms of an object x in this category is just the fundamental group $\pi_1(X, x)$. If we decategorify the fundamental groupoid of X, we forget *how* points in X are connected by paths, remembering only *whether* they are, and we obtain the set of components of X. This captures only the homotopy 0-type of X.

This example shows how decategorification eliminates 'higher-dimensional information' about a situation. Categorification is an attempt to recover this information. This example also suggests that we can keep track of the homotopy 2-type of X if we categorify further and distinguish between paths that are equal and paths that are merely isomorphic (i.e., homotopic). For this we should work with a '2-category' having points of X as objects, paths as morphisms, and certain equivalence classes of homotopies between paths as 2-morphisms.

In a marvelous self-referential twist, the definition of '2-category' is simply the categorification of the definition of 'category'! Like a category, a 2-category has a class of objects, but now for any pair x, y of objects there is no longer a set $\hom(x, y)$; instead, there is a category $\hom(x, y)$. Objects of $\hom(x, y)$ are called morphisms of C, and morphisms between them are called 2-morphisms of C. Composition is no longer a function, but rather a functor:

$$\circ\colon \hom(x, y) \times \hom(y, z) \to \hom(x, z).$$

For any object x there is an identity $1_x \in \hom(x, x)$. And now we have a choice. On the one hand, we can impose associativity and the left and right unit laws strictly, as equational laws. If we do this, we obtain the definition of 'strict 2-category' [**42**]. On the other hand, we can impose them only up to natural isomorphism, with these natural isomorphisms satisfying the coherence laws discussed in the previous section. This is clearly more compatible with the spirit of categorification. If we do this, we obtain the definition of 'weak 2-category' [**12**]. (We warn the reader that strict 2-categories are traditionally known as '2-categories', while weak 2-categories are known as 'bicategories'. The present style of terminology, introduced by Kapranov and Voevodsky [**40**], has the advantage of generalizing easily to n-categories for arbitrary n.)

The classic example of a 2-category is Cat, which has categories as objects, functors as morphisms, and natural transformations as 2-morphisms. The presence of 2-morphisms gives Cat much of its distinctive flavor, which we would miss if we treated it as a mere category. Indeed, Mac Lane has said that categories were originally invented, not to study functors, but to study natural transformations! A good example of two functors that are not equal, but only naturally isomorphic, are the identity functor and the 'double dual' functor on the category of finite-dimensional vector spaces.

Given a topological space X, we can form a 2-category $\Pi_2(X)$ called the 'fundamental 2-groupoid' of X. The objects of this 2-category are the points of X. Given $x, y \in X$, the morphisms from x to y are the paths $f\colon [0,1] \to X$ starting at x and ending at y. Finally, given $f, g \in \hom(x, y)$, the 2-morphisms from f to g are the homotopy classes of paths in $\hom(x, y)$ starting at f and ending at g. Since the associative law for composition of paths holds only up to homotopy, this 2-category is a weak 2-category. One can prove that it captures the homotopy 2-type of X [**18, 47**]. If we decategorify the fundamental 2-groupoid of X, we obtain its fundamental groupoid.

From 2-categories it is a short step to dreaming of n-categories and even ω-categories — but it is not so easy to make these dreams into smoothly functioning mathematical tools. Roughly speaking, an n-category should be some sort of algebraic structure having objects, 1-morphisms between objects, 2-morphisms between 1-morphisms, and so on up to n-morphisms. There should be various ways of composing j-morphisms for $1 \leq j \leq n$, and these should satisfy various laws. As with 2-categories, we can try to impose these laws either strictly or weakly. Strict n-categories have been understood for quite some time now [**23, 28**], but more interesting for us are the weak ones. Various definitions of weak n-category are currently under active study [**5, 10, 36, 57, 58, 61, 62, 63**], and we discuss our own in Section 5. Here, however, we wish to sketch the main challenges any theory of weak n-categories must face, and some of the richness inherent in the notion of weak n-category.

Nota bene: Throughout the rest of this paper, 'n-category' will mean 'weak n-category' unless otherwise specified, and similarly for 'n-groupoid', etc.. For the sake of definiteness, we shall temporarily speak in terms of the 'globular' approach to n-categories developed by Batanin [**10**]. In this approach, for $j \geq 1$ any j-morphism $a\colon f \to g$ has a source f and a target g, which are $(j-1)$-morphisms. When $j \geq 2$ we require that f and g are parallel, by which we mean that $f, g\colon x \to y$ for some $(j-2)$-morphisms x and y. In this approach, we visualize a j-morphism as a j-dimensional disc. For example, for $j = 2$:

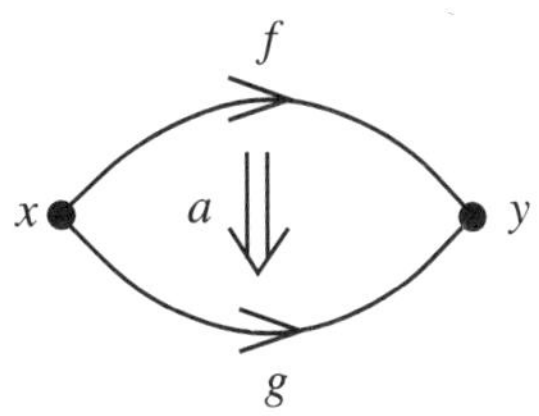

Other approaches to n-categories use j-morphisms with other shapes, such as simplices, discussed in Section 3, or opetopes, discussed in Section 5. We believe that there is basically a single notion of weak n-category lurking behind these different approaches. If this is true, they will eventually be shown to be equivalent, and choosing among them will be merely a matter of convenience. However, the precise meaning of 'equivalence' here is itself rather subtle and n-categorical in flavor [**3, 5**].

The first challenge to any theory of n-categories is to give an adequate treatment of coherence laws. Composition in an n-category should satisfy equational laws only at the top level, between n-morphisms. Any law concerning j-morphisms for $j < n$ should hold only 'up to equivalence'. Here a n-morphism is defined to be an 'equivalence' if it is invertible, while for $j < n$ a j-morphism is recursively defined to be an equivalence if it is invertible *up to equivalence.* Equivalence is generally the correct substitute for the notion of equality in n-categorical mathematics. When laws are formulated as equivalences, these equivalences should in turn satisfy coherence laws of their own, but again only up to equivalence, and so on. This becomes ever more complicated and unmanageable with increasing n unless one takes a systematic approach to coherence laws.

The second challenge to any theory of n-categories is to handle certain key examples. First, for any n, there should be an $(n+1)$-category $n\mathrm{Cat}$, whose objects are (small) n-categories, whose morphisms are suitably weakened functors between these, whose 2-morphisms are suitably weakened natural transformations, and so on. Here by 'suitably weakened' we refer to the fact that all laws should hold only up to equivalence. Second, for any topological space X, there should be an n-category $\Pi_n(X)$ whose objects are points of X, whose morphisms are paths, whose 2-morphisms are paths of paths, and so on, where we take homotopy classes only at the top level. $\Pi_n(X)$ should be an 'n-groupoid', meaning that all its j-morphisms are equivalences for $0 \leq j \leq n$. We call $\Pi_n(X)$ the 'fundamental n-groupoid of X'.

Conversely, any n-groupoid should determine a topological space, its 'geometric realization'. In fact, these constructions should render the study of n-groupoids equivalent to that of homotopy n-types (in a certain sense to be made precise in the next section).

A bit of the richness inherent in the concept of n-category becomes apparent when we make the following observation: an $(n+1)$-category with only one object can be regarded as special sort of n-category. Suppose that C is an $(n+1)$-category with one object x. Then we can form the n-category ΩC by re-indexing: the objects of ΩC are the morphisms of C, the morphisms of ΩC are the 2-morphisms of C, and so on. The n-categories we obtain this way have extra structure. In particular, since the objects of ΩC are really morphisms in C from x to itself, we can 'multiply' (that is, compose) them.

The simplest example is this: if C is a category with a single object x, ΩC is the set of endomorphisms of x. This set is actually a monoid. Conversely, any monoid can be regarded as the monoid of endomorphisms of x for some category with one object x. We summarize this situation by saying that 'a one-object category is a monoid'. Similarly, a one-object 2-category is a monoidal category. It is natural to expect this pattern to continue in all higher dimensions; in fact, it is probably easiest to cheat and *define* a monoidal n-category to be an $(n+1)$-category with one object.

Things get even more interesting when we iterate this process. Given an $(n+k)$-category C with only one object, one morphism, and so on up to one $(k-1)$-morphism, we can form an n-category whose j-morphisms are the $(j+k)$-morphisms of C. In doing so we obtain a particular sort of n-category with extra structure and properties, which we call a 'k-tuply monoidal' n-category. Table 2 shows what we expect these to be like for low values of n and k. For example, the Eckmann-Hilton argument [**4, 8, 27**] shows that a 2-category with one object and one morphism is a commutative monoid. Categorifying this argument, one can show that a 3-category with one object and one morphism is a braided monoidal category. Similarly, we expect that a 4-category with one object, one morphism and one 2-morphism is a symmetric monoidal category, though this has not been worked out in full detail, because of our poor understanding of 4-categories. The fact that both braided and symmetric monoidal categories appear in this table seems to explain why both are natural concepts.

	$n = 0$	$n = 1$	$n = 2$
$k = 0$	sets	categories	2-categories
$k = 1$	monoids	monoidal categories	monoidal 2-categories
$k = 2$	commutative monoids	braided monoidal categories	braided monoidal 2-categories
$k = 3$	‘ ’	symmetric monoidal categories	weakly involutory monoidal 2-categories
$k = 4$	‘ ’	‘ ’	strongly involutory monoidal 2-categories
$k = 5$	‘ ’	‘ ’	‘ ’

2. k-tuply monoidal n-categories

In any reasonable approach to n-categories there should be an $(n+1)$-category $n\mathrm{Cat}_k$ whose objects are k-tuply monoidal weak n-categories. One should also be able to treat $n\mathrm{Cat}_k$ as a full sub-$(n+k)$-category of $(n+k)\mathrm{Cat}$, though even for low n, k this is perhaps not as well known as it should be. Consider for example $n = 0, k = 1$. The objects of $0\mathrm{Cat}_1$ are one-object categories, or monoids. The morphisms of $0\mathrm{Cat}_1$ are functors between one-object categories, or monoid homomorphisms. But $0\mathrm{Cat}_1$ also has 2-morphisms corresponding to natural transformations. We leave it as an exercise to work out what these are in concrete terms. More recently, Kapranov and Voevodsky [**40**] have considered the case $n = k = 1$.

We gave a detailed discussion of Table 2 in an earlier paper [**4**], and subsequent work by various authors has improved our understanding of some of the higher entries [**8, 24, 26**]. There are many interesting processes going from each entry in this table to its neighbors. We list some of the main ones below. Most of these have only been thoroughly studied for low values of n and k, often in the framework of ‘semistrict’ n-categories, which are a kind of halfway house between strict and weak ones. We expect that they all generalize to weak k-tuply monoidal n-categories for arbitrary n and k, but in many cases this has not yet been proved.

- *Decategorification:* $(n, k) \to (n-1, k)$. Let C be a k-tuply monoidal n-category C. Then there should be a k-tuply monoidal $(n-1)$-category $\mathrm{Decat}C$ whose j-morphisms are the same as those of C for $j < n-1$, but whose $(n-1)$-morphisms are isomorphism classes of $(n-1)$-morphisms of C.

- *Discrete categorification:* $(n, k) \to (n+1, k)$. There should be a ‘discrete’ k-tuply monoidal $(n+1)$-category $\mathrm{Disc}C$ having the j-morphisms of C as its j-morphisms for $j \le n$, and only identity $(n+1)$-morphisms. The decategorification of $\mathrm{Disc}C$ should be C.

• *Delooping:* $(n, k) \to (n + 1, k - 1)$. There should be a $(k - 1)$-tuply monoidal $(n + 1)$-category BC with one object obtained by reindexing, the j-morphisms of BC being the $(j + 1)$-morphisms of C. We use the notation 'B' and call BC the 'delooping' of C because of its relation to the classifying space construction in topology.

• *Looping:* $(n, k) \to (n - 1, k + 1)$. Given objects x, y in an n-category, there should be an $(n-1)$-category $\mathrm{hom}(x, y)$. If $x = y$ this should be a monoidal $(n-1)$-category, and we denote it as $\mathrm{end}(x)$. For $k > 0$, if 1 denotes the unit object of the k-tuply monoidal n-category C, $\mathrm{end}(1)$ should be a $(k+1)$-tuply monoidal $(n - 1)$-category. We call this process 'looping', and denote the result as ΩC, because of its relation to loop space construction in topology. For $k > 0$, looping should extend to an $(n+k)$-functor $\Omega\colon n\mathrm{Cat}_k \to (n - 1)\mathrm{Cat}_{k+1}$. The case $k = 0$ is a bit different: we should be able to loop a 'pointed' n-category, one having a distinguished object x, by letting $\Omega C = \mathrm{end}(x)$. In either case, the j-morphisms of ΩC correspond to certain $(j - 1)$-morphisms of C.

• *Forgetting monoidal structure:* $(n, k) \to (n, k-1)$. By forgetting the kth level of monoidal structure, we should be able to think of C as a $(k - 1)$-tuply monoidal n-category FC. This should extend to an n-functor $F\colon n\mathrm{Cat}_k \to n\mathrm{Cat}_{k-1}$.

• *Stabilization:* $(n, k) \to (n, k + 1)$. Though adjoint n-functors are still poorly understood, there should be a left adjoint to forgetting monoidal structure, which we call 'stabilization' and denote by $S\colon n\mathrm{Cat}_k \to n\mathrm{Cat}_{k+1}$. (In our previous work we called it 'suspension', but this is probably a bit misleading.) The 'stabilization hypothesis' [**4**] states that for $k \geq n+2$, stabilization is an equivalence from $n\mathrm{Cat}_k$ to $n\mathrm{Cat}_{k+1}$. This is why the nth column of Table 2 has only $n+2$ distinct entries, and then settles down. While not yet proven or even formulated as a precise conjecture except in low dimensions, there is a lot of good evidence for this hypothesis, some of which we mention in the next section. In what follows, we assume this hypothesis and call a k-tuply monoidal n-category with $k = n + 2$ a 'stable n-category'.

• *Forming the generalized center:* $(n, k) \to (n, k+1)$. Thinking of C as an object of the $(n+k)$-category $n\mathrm{Cat}_k$, there should be a $(k+1)$-tuply monoidal n-category ZC, the 'generalized center' of C, given by $\Omega^k(\mathrm{end}(C))$. In other words, ZC is the largest sub-$(n + k + 1)$-category of $(n + k)\mathrm{Cat}$ having C as its only object, 1_C as its only morphism, 1_{1_C} as its only 2-morphism, and so on up to dimension k. This construction gets its name from the case $n = 0$, $k = 1$, where ZC is the usual center of the monoid C. Categorifying leads to the case $n = 1$, $k = 1$, which gives a very important construction of braided monoidal categories from monoidal categories [**38, 40, 48**]. In particular, when C is the monoidal category of representations of a Hopf algebra H, ZC is the braided monoidal category of representations of the quantum double $D(H)$. Categorifying still further, Baez and Neuchl [**8**] treated the case $n = 2$, $k = 1$. Subsequently Crans [**24**] corrected some errors in their work and dealt with the cases $n = 2$, $k > 1$.

3. Lessons from Homotopy Theory

In Grothendieck's famous 600-page letter to Quillen [**35**], he proposed developing n-category theory as a vast generalization of homotopy theory, with a special class of n-categories — the n-groupoids — corresponding to homotopy n-types. When this idea is finally worked out, we will be able to translate all of homotopy theory into the language of n-groupoids. Eventually this should deepen our understanding of the conceptual foundations of homotopy theory, and help us apply its techniques to other branches of mathematics. But even now, with n-category theory still in its squalling infancy, this translation project is worthwhile. The reason is that homotopy theory is our best source of insight into n-categories. We need to be careful here, since homotopy theory avoids precisely what is most new and interesting about general n-categories, namely the presence of j-morphisms that are not equivalences. However, this is a bit less of a drawback than it might at first seem. After all, the most mysterious aspect of n-category theory is the origin of coherence laws, and these, being implemented as equivalences, appear already in the context of n-groupoids.

ω-groupoids	homotopy types
n-groupoids	homotopy n-types
k-tuply groupal ω-groupoids	homotopy types of k-fold loop spaces
k-tuply groupal n-groupoids	homotopy n-types of k-fold loop spaces
k-tuply monoidal ω-groupoids	homotopy types of E_k spaces
k-tuply monoidal n-groupoids	homotopy n-types of E_k spaces
stable ω-groupoids	homotopy types of infinite loop spaces
stable n-groupoids	homotopy n-types of infinite loop spaces
$\mathbb{Z}$-groupoids	homotopy types of spectra

3. Translating between n-groupoid theory and homotopy theory

We can begin this translation project with the help of Table 3. In explaining this table, we shall use the simplicial approach to n-groupoids instead of the globular approach. The reason is that simplicial methods are quite popular among topologists, so all the necessary machinery has already been developed [**49**]. For other approaches to n-groupoids, see the work of Brown and his collaborators [**16**], Tamsamani [**63**], and Batanin [**10**].

In topology one usually speaks of 'j-cells' rather than j-morphisms. A 'simplicial set' has a set of j-cells for each $j \geq 0$, which we visualize as being shaped like j-simplices. For each j-cell f there are certain $(j-1)$-cells $d_0 f, \ldots, d_n f$ called 'faces' and $(j+1)$-cells $i_0 f, \ldots, i_{n+1} f$ called 'degeneracies'. One also requires that the face and degeneracy maps satisfy certain well-known relations; we will not need the formulas here.

A 'j-dimensional horn' in a simplicial set is, roughly speaking, a configuration in which all but one of the faces of a j-simplex have been filled in by $(j-1)$-cells in a consistent way. A simplicial set for which any horn can be extended to a j-cell is called a 'Kan complex'. A Kan complex is the simplicial version of an ω-groupoid: a structure like an n-groupoid, but without any cutoff on the dimension of the j-morphisms.

To see how this idea works, suppose we have a Kan complex containing a 'composable' pair of 1-cells f and g, meaning that $d_1 f = d_0 g$. This gives a 2-dimensional horn with f and g as two of its faces, so we can extend this horn to a 2-cell F, which has as its third face some 1-cell h:

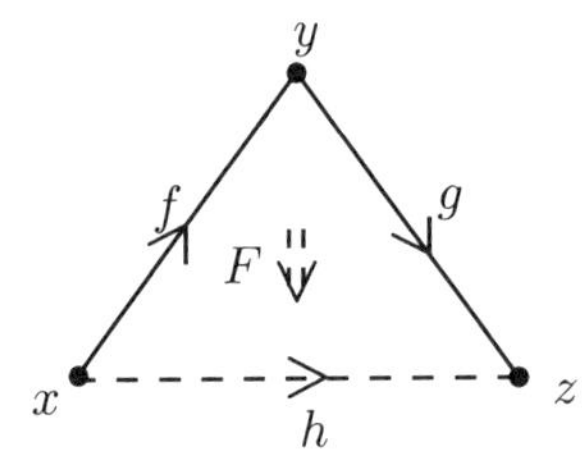

In this situation, we call h 'a composite' of f and g, and think of F as the 'process of composition'. Note that there is not a unique preferred composite, so composition is not an operation in the traditional sense. However, any two composites can be seen to be equivalent, where two j-cells with all the same faces are said to be 'equivalent' if there is a $(j+1)$-cell having them as two of its faces, the rest being degenerate. From an algebraic viewpoint, the reason is that we have defined composition by a *universal property*.

Thanks to the magic of universal properties, Kan complexes are a wonderfully efficient formalism for studying ω-groupoids. In particular, there is no need to explicitly list coherence laws! They are all implicit in the fact that every horn can be extended to a cell, and they automatically become explicit if we arbitrarily choose processes of composition. For example, given a composable triple of 1-cells, one obtains the associator by considering a 3-simplex with these as three of its edges and making clever use of the horn-filling condition. Likewise, given a composable quadruple of 1-cells, one obtains the pentagon coherence law by considering a suitable 4-simplex. In fact, all the higher coherence laws for associativity, which Stasheff [**59**] organized into polyhedra called 'associahedra', have been obtained from higher-dimensional simplices by Street [**61**] in his simplicial approach to ω-categories.

If we take the liberty of calling Kan complexes 'ω-groupoids', we can set up a correspondence between ω-groupoids and homotopy types as follows. Given a topological space X, we can form an ω-groupoid $\Pi(X)$ whose j-cells are all the continuous maps from the standard j-simplex into X, with faces and degeneracies defined in the obvious way. We think of this as the 'fundamental ω-groupoid of X'. Conversely, given an ω-groupoid G, we can form a topological space by taking one geometrical j-simplex for each j-cell of G and gluing them all together using the face and degeneracy maps in the obvious way. This is called the 'geometric realization of G' and denoted $|G|$.

We thus obtain functors going both ways between the category ωGpd, having Kan complexes as objects and simplicial maps between these as morphisms, and the category Top, having nice spaces as objects and continuous maps as morphisms. (We say a space is 'nice' if it is compactly generated and homotopy equivalent to a CW complex; we use this nonstandard definition of Top to exclude various pathologies.) While these functors are adjoint to one another, they do not set up an equivalence of categories. Nonetheless, we expect that ωGpd and Top are 'the same' in a subtler sense — namely, as ω-categories. More precisely, these categories should extend to ω-categories, where the 2-morphisms correspond to homotopies between maps, the 3-morphisms correspond to homotopies between homotopies, and so on. The functors

$$\Pi\colon \mathrm{Top} \to \omega\mathrm{Gpd},$$

$$|\cdot|\colon \omega\mathrm{Gpd} \to \mathrm{Top}$$

should then extend to ω-functors, giving an equivalence of ω-categories.

In the absence of a good theory of ω-categories, topologists have traditionally used other language to express the fact that ωGpd and Top are 'the same' for the purposes of homotopy theory. For example, Π and $|\cdot|$ establish an equivalence between the homotopy category of ω-groupoids and the homotopy category of nice spaces. Here the 'homotopy category' is formed by adjoining formal inverses to all maps inducing isomorphisms of homotopy groups [**32**], where one defines the homotopy groups of a Kan complex to be those of its geometric realization. An object in the homotopy category of Top is called a 'homotopy type' [**11**].

Starting from the correspondence between ω-groupoids and homotopy types we can set up the other correspondences listed in Table 3. For example, we can define an 'n-groupoid' to be a Kan complex such that for $j > n+1$ any configuration in which all the faces of a j-simplex have been filled in by $(j-1)$-cells in a consistent way can be uniquely extended to a j-cell. This ensures that all cells of dimension higher than n play the role of equations. The geometric realization of an n-groupoid is a space with vanishing homotopy groups above dimension n, and the homotopy category of such spaces is called the category of 'homotopy n-types'. The homotopy category of n-groupoids is equivalent to the category of homotopy n-types, and in fact one expects an equivalence of $(n+1)$-categories.

We define a 'k-tuply groupal ω-groupoid' to be a Kan complex with only one j-cell for $j < k$. Under the correspondence between ω-groupoids and homotopy types, these correspond to homotopy types with vanishing homotopy groups below dimension k. Similarly, we define a 'k-tuply groupal n-groupoid' to be an $(n+k)$-groupoid with only one j-cell for $j < k$. These correspond to homotopy $(n+k)$-types with vanishing homotopy groups below dimension k.

Recall from the previous section that we can think of an n-category C with only one j-morphism for $j < k$ as an n-category with extra structure and properties. In the globular approach we do this by 'reindexing', constructing an n-category whose j-morphisms are the $(j+k)$-morphisms of C. Similarly, it is often useful to think of a k-tuply groupal n-groupoid G as an n-groupoid with extra structure and properties. However, in the simplicial approach we cannot simply reindex the cells of G. Instead, we can use a standard construction [**49**] to form a Kan complex $\Omega^k G$ whose geometric realization is homotopy equivalent to the kth loop space of the geometric realization of G. This is why in Table 3 we say that k-tuply groupal n-groupoids correspond to homotopy n-types of k-fold loop spaces. Similarly, k-tuply groupal ω-groupoids correspond to homotopy types of k-fold loop spaces.

Exactly what extra structure and properties does an ω-groupoid have if it is k-tuply groupal? In other words, what extra structure and properties does a space have if it is homotopy equivalent to a k-fold loop space? This question has inspired the development of many interesting mathematical tools. Here we shall describe just one of these, the 'little k-cubes operad', invented by Boardman and Vogt [**13, 14**] and cast into the language of operads by May [**50**].

We begin with the definition of an 'operad'. For each $\ell \geq 0$, an operad O has a set O_ℓ of 'ℓ-ary operations'. We visualize such an operation as a tree with one vertex or 'node', ℓ edges representing inputs coming in from above, and one edge representing the output coming out from below:

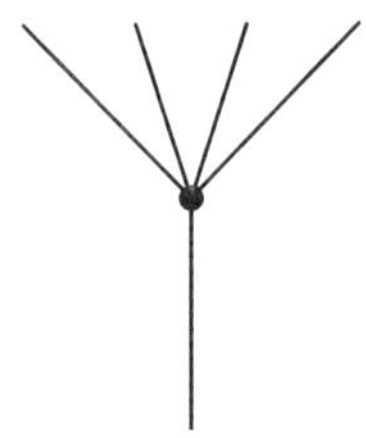

We can compose these trees by feeding the outputs of ℓ of them into one with ℓ inputs:

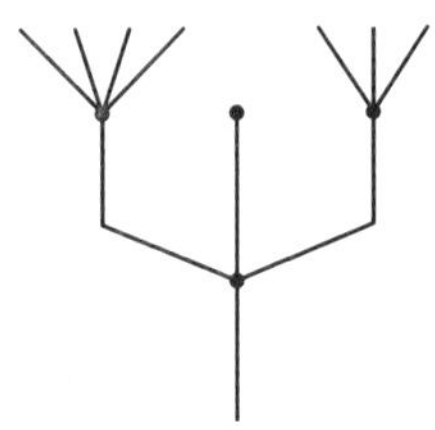

In other words, for any $i_1, \ldots, i_\ell$ there is a function

$$\begin{array}{rcl} O_\ell \times O_{i_1} \times \cdots \times O_{i_\ell} & \to & O_{i_1 + \cdots + i_\ell}. \\ (f, g_1, \ldots, g_\ell) & \mapsto & f \cdot (g_1, \ldots, g_\ell) \end{array}$$

We require that composition is 'associative', meaning that

$$f \cdot (g_1 \cdot (h_{11}, \ldots, h_{1i_1}), \ldots, g_\ell \cdot (h_{\ell 1}, \ldots, h_{\ell i_\ell})) =$$

$$(f \cdot (g_1, \ldots g_\ell)) \cdot (h_{11}, \ldots, h_{1i_1}, \ldots \ldots, h_{\ell 1}, \ldots, h_{\ell i_\ell})$$

whenever both sides are well-defined. This makes composites such as the following one unambiguous:

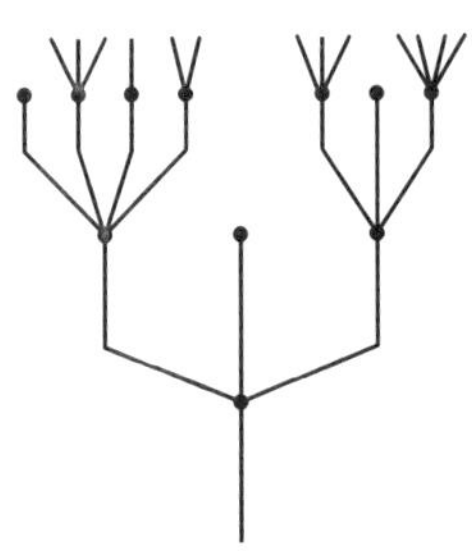

We also require the existence of an unit operation $1 \in O_1$ such that

$$1 \cdot (f) = f, \qquad f \cdot (1, \ldots, 1) = f$$

for all $f \in O_\ell$.

What we have so far is an *planar operad.* For a full-fledged operad, we also assume that there are right actions of the symmetric groups S_ℓ on the sets O_ℓ for which the following compatibility conditions hold. First, for any $f \in O_\ell$, $\sigma \in S_\ell$, and $g_j \in O_{i_j}$ for $1 \leq j \leq \ell$, we require

$$(f\sigma) \cdot (g_{\sigma(1)}, \dots, g_{\sigma(\ell)}) = (f \cdot (g_1, \dots, g_\ell))\, \rho(\sigma),$$

where

$$\rho\colon S_\ell \to S_{i_1 + \cdots + i_\ell}$$

is the obvious homomorphism. Second, for any $f \in O_\ell$, and $g_j \in O_{i_j}$, $\sigma_j \in S_{i_j}$ for $1 \leq j \leq \ell$, we require

$$f \cdot (g_1\sigma_1, \dots, g_\ell\sigma_\ell) = (f \cdot (g_1, \dots, g_\ell))\, \rho'(\sigma_1, \dots, \sigma_\ell),$$

where

$$\rho'\colon S_{i_1} \times \cdots \times S_{i_\ell} \to S_{i_1 + \cdots + i_\ell}$$

is the obvious homomorphism.

Just as groups are interesting for their actions, operads are interesting for their 'algebras'. Given an operad O, an 'O-algebra' is a set A equipped with actions

$$\alpha\colon O_\ell \times A^\ell \to A,$$

or equivalently, maps

$$\alpha\colon O_\ell \to \hom(A^\ell, A)$$

representing the ℓ-ary operations of O as actual operations on A. We require that α sends the identity operation $1 \in O_1$ to the identity function on A and sends composites to composites:

$$\alpha(f \cdot (g_1, \dots, g_\ell)) = \alpha(f) \circ (\alpha(g_1) \times \cdots \times \alpha(g_\ell)).$$

We also require that

$$\alpha(f\sigma) = \alpha(f)\sigma,$$

where $f \in O_\ell$, and $\sigma \in S_\ell$ acts on $\hom(A^\ell, A)$ on the right by permuting the factors in A^ℓ. We omit this requirement if O is merely planar.

More generally, one can define operads and their algebras in any symmetric monoidal category C, which amounts to replacing sets and functions in the above definitions by objects and morphisms in C, and replacing the Cartesian product by the tensor product in C. We shall mainly be interested in operads in the category Top, which are called 'topological operads'. Spaces with extra structure and properties can often be described as algebras of topological operads.

The most interesting example for us is the 'little k-cubes operad', $C(k)$. Here the space $C(k)_\ell$ of ℓ-ary operations is the space of ℓ disjoint k-cubes linearly embedded in the cube $[0,1]^k$ with their edges parallel to the coordinate axes:

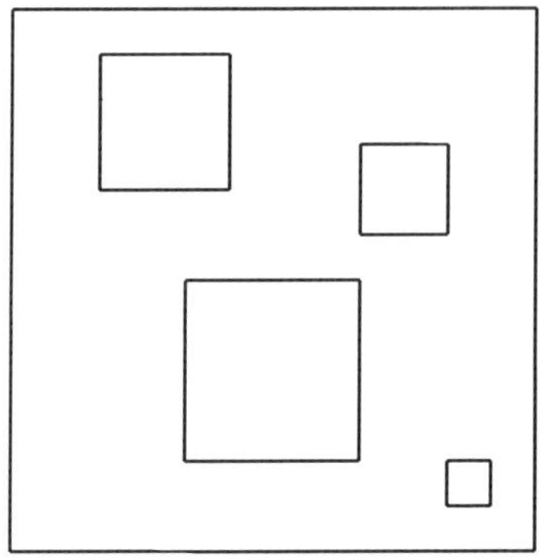

An algebra of the little k-cubes operad is called an 'E_k space'. Given a space X with a basepoint, there is an obvious way to make $\Omega^k X$ into an E_k space using the fact that it consists of maps $f\colon [0,1]^k \to X$ sending the boundary of the cube to the basepoint. Conversely, the set of components of an E_k space automatically becomes a monoid, and if this monoid is a group then the E_k space is homotopy equivalent to a k-fold loop space. Homotopy types of E_k spaces are thus a bit more general than homotopy types of k-fold loop spaces. While the latter correspond to k-tuply groupal ω-groupoids, the former should correspond to 'k-tuply monoidal ω-groupoids', that is, ω-categories with only one j-morphism for $j < k$, for which all j-morphisms are equivalences for $j > k$ — but not necessarily for $j = k$. There should similarly be a correspondence between k-tuply monoidal n-groupoids and homotopy n-types of E_k spaces.

At the end of the previous section we listed various processes going between neighboring entries of Table 2. When restricted to k-tuply groupal n-groupoids, most of these processes have well-known topological interpretations, which we summarize below. We let $n\mathrm{Type}_k$ stand for the category of homotopy n-types of k-fold loop spaces. It is easiest to define this as the full subcategory of the homotopy category of nice spaces X with basepoint such that $\pi_j(X) = 0$ unless $k \le j \le n+k$. However, by repeated looping, we shall think of the objects of $n\mathrm{Type}_k$ as k-fold loop spaces with vanishing homotopy groups above dimension n. A technical point worth noting is that for $k = 0$, these really correspond to n-groupoids *equipped with a distinguished 0-cell.*

- *Decategorification:* $(n,k) \to (n-1,k)$. Let X be a k-fold loop space with $\pi_j(X) = 0$ for $j > n$. Then we can kill off its nth homotopy group by attaching cells, obtaining a k-fold loop space $\mathrm{Decat}X$ with $\pi_j(X) = 0$ for $j > n-1$. There is a map $X \to \mathrm{Decat}X$ inducing isomorphisms $\pi_j(X) \cong \pi_j(\mathrm{Decat}X)$ for $j \le n-1$. This process gives a functor $\mathrm{Decat}\colon n\mathrm{Type}_k \to (n-1)\mathrm{Type}_k$.

- *Discrete categorification:* $(n,k) \to (n+1,k)$.
The forgetful functor $\mathrm{Disc}\colon n\mathrm{Type}_k \to (n+1)\mathrm{Type}_k$ is the left adjoint of Decat.

- *Delooping:* $(n,k) \to (n+1,k-1)$. There is a $(k-1)$-fold loop space BX, called the 'classifying space' or 'delooping' of X, with $\pi_j(BX) \cong \pi_{j-1}(X)$. This process gives a functor $B\colon n\mathrm{Type}_k \to (n+1)\mathrm{Type}_{k-1}$.

• *Looping:* $(n,k) \to (n-1,k+1)$. For $k>0$, there is a $(k+1)$-fold loop space ΩX, consisting of all based loops in X, with $\pi_j(\Omega X) \cong \pi_{j-1}(X)$. This process gives a functor $\Omega\colon n\mathrm{Type}_k \to (n-1)\mathrm{Type}_{k+1}$ that is right adjoint to B. For $k>0$, delooping followed by looping is naturally isomorphic to the identity functor.

• *Forgetting monoidal structure:* $(n,k) \to (n,k-1)$.
There is a forgetful functor $F\colon n\mathrm{Type}_k \to n\mathrm{Type}_{k-1}$.

• *Stabilization:* $(n,k) \to (n,k+1)$. There is a left adjoint to F called 'stabilization', which we denote by $S\colon n\mathrm{Type}_k \to n\mathrm{Type}_{k+1}$. If we define the 'suspension' functor $\Sigma\colon (n-1)\mathrm{Type}_k \to n\mathrm{Type}_k$ to be the left adjoint of the composite

$$n\mathrm{Type}_k \xrightarrow{\ \Omega\ } (n-1)\mathrm{Type}_{k+1} \xrightarrow{\ F\ } (n-1)\mathrm{Type}_k$$

then stabilization is naturally isomorphic to suspension followed by looping. The Freudenthal suspension theorem says that stabilization is an equivalence for $k \geq n+2$.

• *Forming the generalized center:* $(n,k) \to (n,k+1)$. This process needs to be adapted to stay within the world of k-tuply groupal groupoids, or in other words, homotopy n-types of k-fold loop spaces. We let $\mathrm{aut}(X)$ be the automorphism group of X as an object of $n\mathrm{Type}_k$, and define $ZX = \Omega^k(\mathrm{aut}(X))$.

The fact that stabilization is an equivalence for $k \geq n+2$ leads us to define a 'stable n-groupoid' to be an $(n+2)$-tuply groupal n-groupoid. We expect, in fact, that there is an $(n+1)$-category of stable n-groupoids, and that the decategorification of this is the n-category of stable $(n-1)$-groupoids. If we could take the inverse limit in a suitable sense, we would hope to obtain an ω-category of 'stable ω-groupoids'. These should correspond to what topologists call 'infinite loop spaces', an infinite loop space being a sequence of spaces $X_0, X_1, \ldots$, equipped with homeomorphisms $f_k\colon X_k \to \Omega X_{k+1}$.

Infinite loop spaces play an important role in stable homotopy theory [**1, 29, 50**]. A closely related concept is that of a 'spectrum'. Just as an infinite loop space should correspond to a stable ω-groupoid, a spectrum should correspond to a '$\mathbb{Z}$-groupoid', some sort of gadget with j-morphisms for all $j \in \mathbb{Z}$, all of which are equivalences. In fact, strict $\mathbb{Z}$-categories are easily defined in the globular approach, and strict $\mathbb{Z}$-groupoids then work out to be the same as $\mathbb{Z}$-graded chain complexes of abelian groups. Spectra can be viewed as a generalization of such chain complexes. Apart from what we know about spectra, however, the theory of 'weak $\mathbb{Z}$-categories' remains largely *terra incognita.*

4. Examples of Categorification

In what follows, we consider iterated categorifications of the natural numbers, and some variations on this theme where we adjoin formal inverses or duals. We shall see that quite a bit of mathematics amounts to the study of the resulting objects, some of which are astoundingly complicated. The more 'n-groupoidal' examples are familiar from homotopy theory, but the importance of some of the more 'n-categorical' ones has only become clear in recent work on topological quantum field theory.

The natural numbers are the free monoid on one element, while the integers are the free group on one element. Some categorified analogs of these notions are listed in Table 4. In the rest of this section, we work through these examples in detail.

We begin with the simplest algebraic structure of all: sets. The free set on one element is just the one-element set, denoted by 1. Note that here we are using the word 'the' in a generalized sense. Since all sets with the same cardinality are isomorphic, any singleton $\{x\}$ can be regarded as 'the' 1-element set. In a set, when we speak of 'the' element with some property, we imply that any other element having this property is *equal* to this one, but in a category, when we speak of 'the' object with some property, we mean that any other object having this property is *isomorphic* to this one — typically by means of a uniquely specified isomorphism. More generally, when we speak of 'the' object of an n-category having some property, we mean that any other object with this property is *equivalent* — typically by means of an equivalence which is specified uniquely up to an equivalence which is specified uniquely up to ... and so on. This recursive weakening of the notion of uniqueness, and therefore of the meaning of 'the', is fundamental to categorification.

sets	1
monoids	$\mathbb{N}$
groups	$\mathbb{Z}$
k-tuply monoidal n-categories	$n\mathrm{Braid}_k$
k-tuply monoidal ω-categories	Braid_k
stable n-categories	$n\mathrm{Braid}$
stable ω-categories	Braid_∞
k-tuply monoidal n-categories with duals	$n\mathrm{Tang}_k$
stable n-categories with duals	$n\mathrm{Cob}$
k-tuply groupal n-groupoids	$\Pi_n(\Omega^k S^k)$
k-tuply groupal ω-groupoids	$\Pi(\Omega^k S^k)$
stable ω-groupoids	$\Pi(\Omega^\infty S^\infty)$
$\mathbb{Z}$-groupoids	the sphere spectrum

4. Algebraic structures and the free such structures on one generator

Next consider the free category on one object. This is just the category with one object x and one morphism $1_x\colon x \to x$. More generally, for each n, consider the free n-category on one object. In the globular approach to n-categories we may take this to be the n-category with one object x, one morphism $1_x\colon x \to x$, one 2-morphism $1_{1_x}\colon 1_x \Rightarrow 1_x$, and so on up to one n-morphism. This is an n-groupoid, namely the globular version of the fundamental n-groupoid of a point. By the remarks in the previous paragraph, the fundamental n-groupoid of any contractible space may also be considered 'the free n-category on one object'.

Things become more interesting when we generalize further and consider the free k-tuply monoidal n-category on one object, which we denote by $n\mathrm{Braid}_k$, for reasons soon to be apparent. Let us see what this looks like in the simplest cases, namely $n = 0$ and 1. The stabilization hypothesis says that we only need to consider $k \leq n + 2$.

- $0\mathrm{Braid}_0$, the free set on one element. This is the one-element set, 1.

- $0\mathrm{Braid}_1$, the free monoid on one element. This is the natural numbers, $\mathbb{N}$, with addition as its monoid structure.

- $0\mathrm{Braid}_2$, the free commutative monoid on one element. This is again $\mathbb{N}$, now regarded as a commutative monoid with addition as its monoid structure.

- $1\mathrm{Braid}_0$, the free category on one object. This is the category with one object and one morphism.

- $1\mathrm{Braid}_1$, the free monoidal category on one object x. The objects of this category are the tensor powers $x^{\otimes \ell}$, and the only morphisms are identity morphisms. (Here we are using the Mac Lane coherence theorem to make $1\mathrm{Braid}_1$ into a strict monoidal category.) This is the discrete categorification of $0\mathrm{Braid}_1$.

- $1\mathrm{Braid}_2$, the free braided monoidal category on one object x. This is the braid groupoid. The objects of this groupoid are the tensor powers $x^{\otimes \ell}$, and the only morphisms are automorphisms, with $\mathrm{end}(x^{\otimes \ell})$ being the ℓ-strand braid group B_ℓ, which has generators σ_i $(1 \leq i \leq \ell - 1)$ and relations

$$\begin{aligned} \sigma_i \sigma_{i+1} \sigma_i &= \sigma_{i+1} \sigma_i \sigma_{i+1}, \\ \sigma_i \sigma_j &= \sigma_j \sigma_i \qquad\qquad \text{for } |i - j| > 1. \end{aligned}$$

- $1\mathrm{Braid}_3$, the free symmetric monoidal category on one object x. This is the symmetric groupoid. The objects of this groupoid are the tensor powers $x^{\otimes \ell}$, and the only morphisms are automorphisms, with $\mathrm{end}(x^{\otimes \ell})$ being the symmetric group on ℓ letters, S_ℓ. This group can be given a presentation like that of B_ℓ, but with the additional relations

$$\sigma_i^2 = 1.$$

The symmetric groupoid is equivalent to the category with finite sets as objects and bijections as morphisms. Note that, like the category FinSet having finite sets as objects and arbitrary functions as morphisms, this category is a rig category with $\mathbb{N}$ as its decategorification.

The following heuristic argument allows us to guess a general formula for $n\mathrm{Braid}_k$. In all the cases considered above, $n\mathrm{Braid}_k$ is a k-tuply monoidal n-groupoid. This should hold in general, since for $j > 0$ the j-morphisms of $n\mathrm{Braid}_k$ arise from coherence laws, and should thus be equivalences. We also expect that $n\mathrm{Braid}_k$ is the 'free k-tuply monoidal n-groupoid on one object'. By Table 3,

$n\mathrm{Braid}_k$ should thus be the fundamental n-groupoid of some E_k space $X_{n,k}$ with vanishing homotopy groups above dimension n. We also expect that $\mathrm{Decat}((n+1)\mathrm{Braid}_k) \simeq n\mathrm{Braid}_k$, so there should be some sort of inverse limit, the 'free k-tuply monoidal ω-groupoid on one object', which we denote by Braid_k. Corresponding to this there should be a sequence of maps $X_{n+1,k} \to X_{n,k}$, with $X_{n,k}$ obtained from $X_{n+1,k}$ by killing its $(n+1)$st homotopy group, and with the inverse limit being a space X_k whose fundamental ω-groupoid is Braid_k. By the correspondence between k-tuply monoidal ω-groupoids and E_k spaces, we expect that X_k is the 'free E_k space on one point'.

While this argument involves many forms of reasoning that have not yet been made rigorous, there is at least a precise meaning to the 'free E_k space on one point'. Given any operad O in the category of pointed spaces and any pointed space X, there is a standard construction of the 'free O-algebra on X', due to May [**50**]. To form the free O-algebra on one point, we should first form the free free pointed space on one point, namely S^0, and then apply this standard construction. The result is the disjoint union

$$\coprod_{\ell=0}^{\infty} O_\ell/S_\ell$$

which becomes an O-algebra in a tautologous way.

In the case at hand, since the the operad for E_k spaces is the little k-cubes operad $C(k)$, the free E_k space on one point is

$$X_k = \coprod_{\ell=0}^{\infty} C(k)_\ell/S_\ell.$$

We expect, therefore, an equivalence

$$n\mathrm{Braid}_k \simeq \Pi_n(X_k).$$

To work with this equivalence, it is helpful to note that $C(k)_\ell$ is homotopy equivalent to the 'configuration space' [**55**] of ℓ distinct points in the k-cube:

$$\{(x_1, \ldots, x_\ell) \in [0,1]^k \; : \; x_i \neq x_j \text{ if } i \neq j\}.$$

Moreover, this homotopy equivalence is compatible with the obvious actions of S_ℓ. It follows that X_k is homotopy equivalent to the pointed space of all finite sets of distinct points in the k-cube, where the empty set plays the role of basepoint.

To see how this works in an example, consider the case $n = 1$, $k = 2$. The space X_2 is equivalent to the space of finite sets of points in the square. A path in X_2 amounts to a braid with an arbitrary number of strands. It follows that the fundamental groupoid of X_2 is equivalent to the braid groupoid, so

$$1\mathrm{Braid}_2 \simeq \Pi_1(X_2).$$

More generally, in the globular approach to n-categories, the n-morphisms of $n\mathrm{Braid}_k$ should correspond to certain n-dimensional surfaces in $[0,1]^{n+k}$, which we could call 'n-braids in $n+k$ dimensions'. We see here an instance of a general theme, namely that in a k-tuply monoidal n-category the number n often plays the role of 'dimension', while k plays the role of 'codimension'.

Now let us turn to the 'free k-tuply groupal n-groupoid on one object', which we temporarily denote by $G_{n,k}$. For $n = 0$, $k = 1$ this is just the free group on one element, namely the integers, $\mathbb{Z}$. For higher values of n and k we may thus regard

$G_{n,k}$ as a categorified, stabilized version of $\mathbb{Z}$. In what follows we restrict attention to the case $k > 0$, since in this case a k-tuply groupal n-groupoid automatically has a distinguished object, the unit.

Again we can use a heuristic argument to guess a formula for $G_{n,k}$. Since we expect that $G_{n,k} \simeq \mathrm{Decat} G_{n+1,k}$, there should be some sort of inverse limit G_k, the 'free k-tuply groupal ω-groupoid on one object'. By Table 3 we expect G_k to be the fundamental ω-groupoid of some k-fold loop space, so the problem is to determine this space. Just as the group $\mathbb{Z}$ is obtained from the monoid $\mathbb{N}$ by adjoining formal inverses, G_k should be obtained from Braid_k by adjoining formal weak inverses for all objects. More generally, we should be able to turn any k-tuply monoidal ω-groupoid into a k-tuply groupal ω-groupoid by adjoining formal weak inverses of objects. In the language of homotopy theory, this process should turn E_k spaces into k-fold loop spaces. In fact, this process is familiar in homotopy theory under the name of 'group completion' [**9**]. Since $n\mathrm{Braid}_k$ is the fundamental ω-groupoid of X_k, we thus expect G_k to be the fundamental ω-groupoid of the group completion of X_k. The group completion of X_k is homotopy equivalent to $\Omega^k S^k$, so we expect an equivalence

$$G_{n,k} \simeq \Pi_n(\Omega^k S^k).$$

Group completion automatically gives a map from X_k to $\Omega^k S^k$ which induces a k-tuply monoidal n-functor

$$\Pi_n(X_k) \to \Pi_n(\Omega^k S^k).$$

If the above guesses are correct, n-braids and the homotopy types of spheres play a fundamental role in n-category theory. Since the homotopy groups of spheres are notoriously hard to compute, this means that n-category theory has a certain built-in complexity. Perhaps we should amplify on this a bit. Suppose that C is a globular $(n+k)$-category and x is any object of C. Let $1_1 = 1_x$ and recursively define 1_{i+1} to be 1_{1_i}. Then we are claiming that any k-morphism $f\colon 1_{k-1} \to 1_{k-1}$ determines, at least up to equivalence, a k-tuply monoidal n-functor

$$n\mathrm{Braid}_k \to \mathrm{end}(1_{k-1})$$

mapping the generator of $n\mathrm{Braid}_k$ to f. Moreover, if f is an equivalence, we claim this factors through a k-tuply groupal n-functor

$$\Pi_n(\Omega^k S^k) \to \mathrm{end}(1_{k-1})$$

whose range consists entirely of equivalences.

In our previous exploration of these ideas [**4**], we emphasized the importance of a notion lying halfway between n-groupoids and fully general n-categories, which we called 'n-categories with duals'. The idea here is that duals are an interesting generalization of inverses. In particular, the 'tangle hypothesis' states that there is a 'free k-tuply monoidal n-category with duals on one object', $n\mathrm{Tang}_k$, having as n-morphisms certain n-dimensional surfaces in $[0,1]^{n+k}$ called 'framed n-tangles in $n+k$ dimensions'. For $n=1, k=2$ this was proved by Freyd and Yetter, Turaev, and Shum [**30, 56, 65, 66**]. This special case serves as the basis of recent work on 3-dimensional topological quantum field theory. Indeed, the fact that n-categories with duals are more general than n-groupoids is the reason why topological quantum field theory can give more refined information than homotopy theory. Recently progress has been made on the case $n = 2$, $k = 2$, which has also illuminated the theory of 2-braids in 4 dimensions [**6, 7, 25, 43**]. In general, we expect that in the stable range $n\mathrm{Tang}_k$ is equivalent to the stable n-category of 'framed cobordisms'. Also, the universal property of $n\mathrm{Tang}_k$ should give a k-tuply monoidal n-functor

$$T\colon n\mathrm{Tang}_k \to \Pi_n(\Omega^k S^k),$$

generalizing the Thom-Pontryagin construction. For more details the reader must turn to the references. Our main point here is that if the tangle hypothesis holds, a great deal, not only of homotopy theory, but also of topological quantum field theory arises naturally from the study of categorified analogs of $\mathbb{Z}$!

To conclude, let us note that all the entries in Table 4 should be equipped with 'multiplication' as well as 'addition' operations. For example, by virtue of being the free monoid on one element, $\mathbb{N}$ automatically becomes a rig in the following way: given an element $n \in \mathbb{N}$, there is a unique monoid homomorphism $f\colon \mathbb{N} \to \mathbb{N}$ with $f(1) = n$, namely multiplication by n. Likewise, by virtue of being the free group on one element, $\mathbb{Z}$ automatically becomes a ring. Categorifying once and stabilizing various numbers of times, we see that for $k > 0$, $1\mathrm{Braid}_k$ is a rig category and $\Pi_1(\Omega^k S^k)$ is a 'ring category': a rig category for which objects have additive inverses. This pattern should continue throughout the rest of Table 4.

For example, in stable homotopy theory it is well known that the sphere spectrum is a 'ring spectrum' [**51**]. We expect that the sphere spectrum corresponds to the 'free $\mathbb{Z}$-groupoid on one object'. Indeed, as Joyal once remarked to us, the sphere spectrum could be called 'the true integers', since it is an infinitely categorified, infinitely stabilized analog of $\mathbb{Z}$.

5. n-Categories and the Algebra of Opetopes

We hope the previous sections have whetted the reader's appetite for a rigorous theory of n-categories by sketching a bit of what we might do if we had one. Now we turn to the issue of actually developing this theory. It seems that any definition of n-category involves a choice of the basic shapes of j-morphisms — globes, simplices, or whatever. It also involves a choice of ways to compose j-morphisms by gluing these basic shapes together. Most importantly, it requires a careful treatment of coherence laws. In what follows we present an approach in which all these issues are handled simultaneously using the formalism of operads. In this approach, the basic shapes of j-morphisms are the j-dimensional 'opetopes'. The allowed ways of composing j-morphisms correspond precisely to the $(j+1)$-dimensional opetopes. Moreover, the sequence of higher coherence laws satisfied by composition correspond to opetopes of ever higher dimension.

Before going into the details, let us sketch how this works. First consider some low-dimensional opetopes. The only 0-dimensional opetope is the point:

The only way to glue together 0-dimensional opetopes is the trivial way: the identity operation. The only 1-dimensional opetope is thus the interval, or more precisely the arrow:

The allowed ways of gluing together 1-dimensional opetopes are given by the 2-dimensional opetopes. The first few 2-dimensional opetopes are as follows:

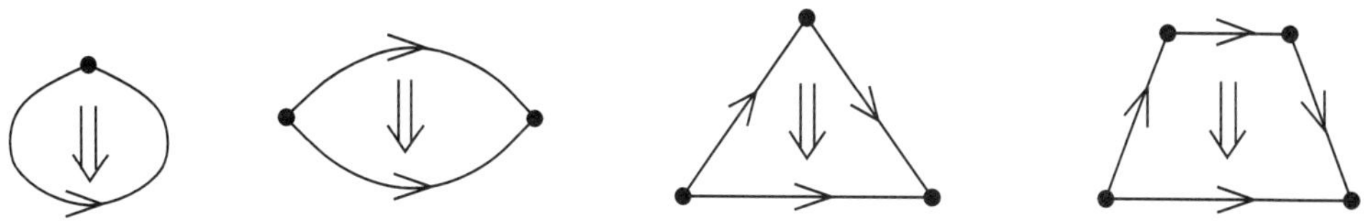

The allowed ways of gluing together 2-dimensional opetopes are given by the 3-dimensional opetopes. There are many of these; a simple example is as follows:

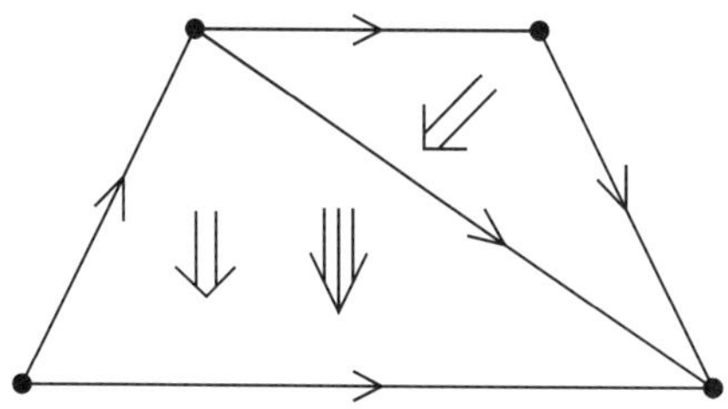

This may be a bit hard to visualize, but it depicts a 3-dimensional shape whose front consists of two 3-sided 'infaces', and whose back consists of a single 4-sided 'outface'. We have drawn double arrows on the infaces but not on the outface. Note that while this shape is topologically a ball, it cannot be realized as a polyhedron with planar faces. This is typical of opetopes.

In general, an $(n + 1)$-dimensional opetope has any number of infaces and exactly one outface: the infaces are n-dimensional opetopes glued together in a tree-like pattern, while the outface is a single n-dimensional opetope. For example, the 3-dimensional opetope above corresponds to the following tree:

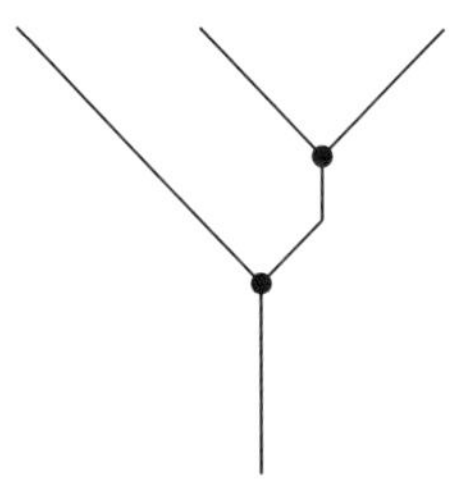

The two triangular infaces of the opetope correspond to the two nodes in this tree. This is a rather special tree; in general, we allow nonplanar trees with any number of nodes and any number of edges coming into each node.

Our approach to n-categories is a bit like the Kan complex approach to n-groupoids described in Section 3, but with simplicial sets replaced by 'opetopic sets'. Basically, an opetopic set is a set of 'cells' shaped like opetopes, such that any face of a cell is again a cell. In an n-category, the j-dimensional cells play the role of j-morphisms. An opetopic set is an n-category if it satisfies the following two properties:

1) *"Any niche has a universal occupant."* A 'niche' is a configuration where all the infaces of an opetope have been filled in by cells, but not the outface or the opetope itself:

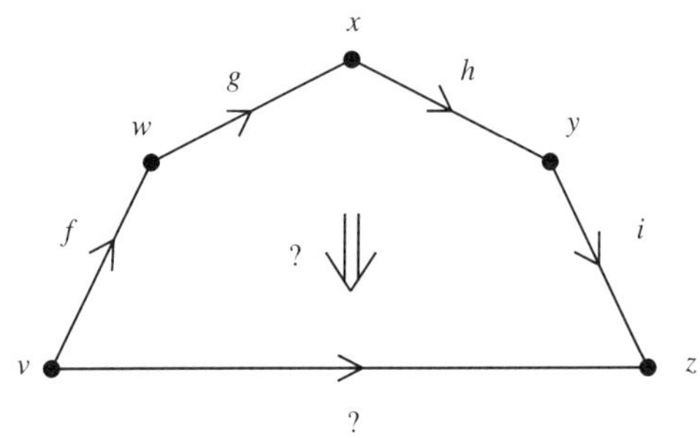

An 'occupant' of the niche is a way of extending this configuration by filling in the opetope (and thus its ouface) with a cell:

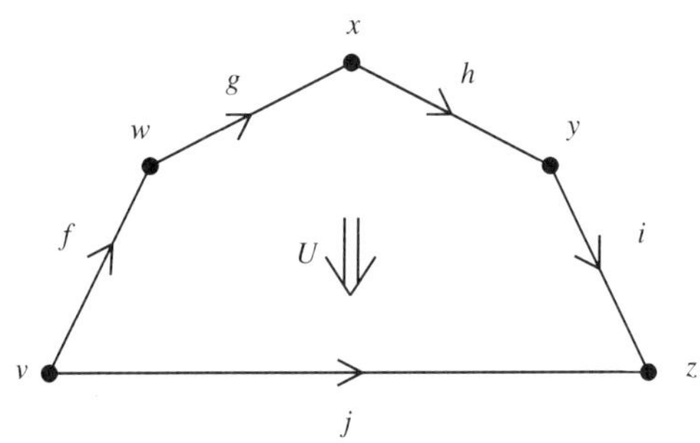

The 'universality' of an occupant means roughly that every other occupant factors through the given one *up to equivalence.* To make this precise we need to define universality in a rather subtle recursive way. We may think of a universal occupant of a niche as 'a process of composing' the infaces, and its outface as 'a composite' of the infaces.

2) *"Composites of universal cells are universal."* Suppose that U, V, and W below are universal cells:

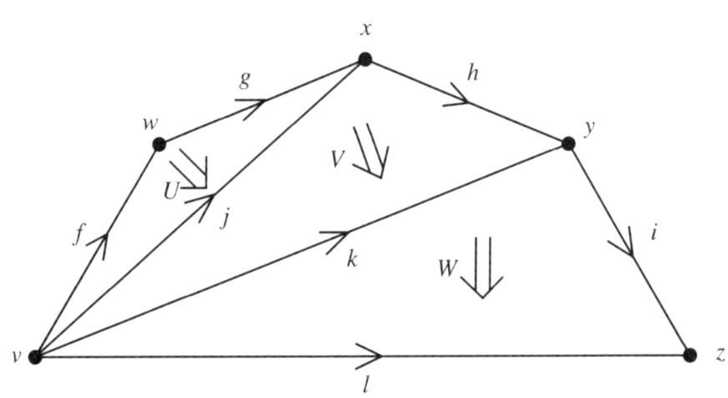

Then we can compose them, and we are guaranteed that their composite is again universal, and thus that the outface l is a composite of the cells f, g, h, i. Note that a process of composing U, V, W is described by a universal occupant of a niche of one higher dimension.

Note that in this approach to n-categories, composition of cells is not an operation in the traditional sense: the composite is defined by a universal property, and is thus unique only up to equivalence. Only at the top level, for the n-cells of an n-category, is the composite truly unique. The main advantage of defining composition by a universal property is that we do not need to list coherence laws:

all the right coherence laws arise automatically! This is a very important point, because in some sense it answers the puzzle concerning the origin of coherence laws.

At first this answer may seem as puzzling as the puzzle it answers. *Why* does defining composition by a universal property automatically generate all the right coherence laws? One reason is that coherence laws are 'right' when they hold in interesting examples, and in these examples composition is usually defined by a universal property. Consider for example the categorified version of $\mathbb{N}$ discussed in the Introduction: the category FinSet. Corresponding to addition in $\mathbb{N}$, the category FinSet has finite coproducts, i.e., disjoint unions. Coproducts are defined by a universal property, and this universal property immediately implies number of things. First, coproducts are unique up to canonical isomorphism. Second, if we pick a coproduct $x \sqcup y$ for every pair of objects $x, y \in$ FinSet, making disjoint union into an operation in the traditional sense, we obtain natural isomorphisms

$$a_{x,y,z} \colon (x \sqcup y) \sqcup z \to x \sqcup (y \sqcup z),$$

$$l_x \colon \emptyset \sqcup x \to x, \qquad r_x \colon x \sqcup \emptyset \to x.$$

Third, these natural isomorphisms satisfy coherence laws making FinSet into a monoidal category. In short, the standard definition of monoidal category, which lists coherence laws, is best regarded as a spinoff of the fact that monoidal structures defined by universal properties automatically satisfy these laws.

Studying this example also suggests another idea which is built into our definition of n-categories. To prove the existence of the associator $a_{x,y,z}$ one uses the universal property for the coproduct of *three* objects, and to prove the existence of l_x and r_x one uses the universal property for the coproduct of *one* object. This suggests that in an n-category, arbitrary ℓ-ary composites should be treated on an equal footing with binary composites. The formalism of operads is admirably suited for this task.

In what follows we first review the theory of typed operads. Then we use this to define the opetopes, and more generally, 'O-opetopes' for any typed operad O. After a brief discussion of some notions concerning O-opetopic sets, we give the the definition of n-categories, and more generally, n-coherent O-algebras. We skim over some technical details which can be found in our paper [**5**].

5.1. Typed Operads. To describe 'many-sorted' or 'typed' algebraic structures using operads, we need to generalize the concept of operad a bit. For any set S of 'types', there is a notion of 'S-operad'. The basic idea is that for any $x_1, \ldots, x_\ell, x' \in S$, an S-operad O has a set $O(x_1, \ldots, x_\ell; x')$ of k-ary operations with inputs of type $x_1, \ldots, x_\ell$ and output of type x'. As in an ordinary untyped operad, we can visualize such an operation as a tree, but now we label the edges of the tree by types. For example, an operation $f \in O(x_1, \ldots, x_4; x')$ is drawn as follows:

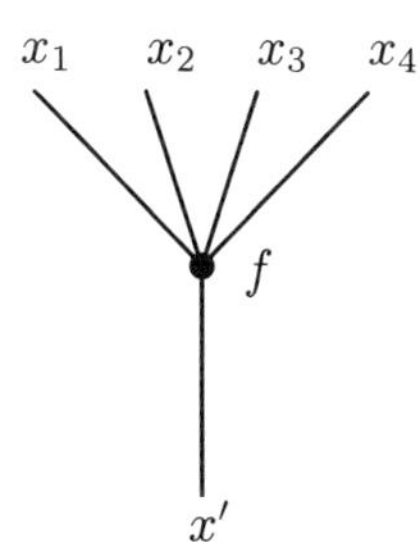

More precisely:

DEFINITION 1. *For any set S, an 'S-operad' O consists of*

1. *for any $x_1, \ldots, x_\ell, x' \in S$, a set $O(x_1, \ldots, x_\ell; x')$*
2. *for any $f \in O(x_1, \ldots, x_\ell; x')$ and any $g_1 \in O(x_{11}, \ldots, x_{1i_1}; x_1)$, $\ldots, g_\ell \in O(x_{\ell 1}, \ldots, x_{\ell i_\ell}; x_\ell)$, an element*
$$f \cdot (g_1, \ldots, g_\ell) \in O(x_{11}, \ldots, x_{1i_1}, \ldots \ldots, x_{\ell 1}, \ldots, x_{\ell i_\ell}; x')$$
3. *for any $x \in S$, an element $1_x \in O(x; x)$*
4. *for any permutation $\sigma \in S_\ell$, a map*
$$\begin{array}{rcl} \sigma\colon O(x_1, \ldots, x_\ell; x') & \to & O(x_{\sigma(1)}, \ldots, x_{\sigma(\ell)}; x') \\ f & \mapsto & f\sigma \end{array}$$

such that:

(a) *whenever both sides make sense,*
$$f \cdot (g_1 \cdot (h_{11}, \ldots, h_{1i_1}), \ldots, g_\ell \cdot (h_{\ell 1}, \ldots, h_{\ell i_\ell})) =$$
$$(f \cdot (g_1, \ldots g_\ell)) \cdot (h_{11}, \ldots, h_{1i_1}, \ldots \ldots, h_{\ell 1}, \ldots, h_{\ell i_\ell})$$
(b) *for any $f \in O(x_1, \ldots, x_\ell; x')$,*
$$f = 1_{x'} \cdot f = f \cdot (1_{x_1}, \ldots, 1_{x_\ell})$$
(c) *for any $f \in O(x_1, \ldots, x_\ell; x')$ and $\sigma, \sigma' \in S_\ell$,*
$$f(\sigma\sigma') = (f\sigma)\sigma'$$
(d) *for any $f \in O(x_1, \ldots, x_\ell; x')$, $\sigma \in S_\ell$, and $g_1 \in O(x_{11}, \ldots, x_{1i_1}; x_1)$, $\ldots, g_\ell \in O(x_{\ell 1}, \ldots, x_{\ell i_\ell}; x_\ell)$,*
$$(f\sigma) \cdot (g_{\sigma(1)}, \ldots, g_{\sigma(\ell)}) = (f \cdot (g_1, \ldots, g_\ell))\, \rho(\sigma),$$
where $\rho\colon S_\ell \to S_{i_1 + \cdots + i_\ell}$ is the obvious homomorphism.
(e) *for any $f \in O(x_1, \ldots, x_\ell; x')$, $g_1 \in O(x_{11}, \ldots, x_{1i_1}; x_1), \ldots,$ $g_\ell \in O(x_{k1}, \ldots, x_{\ell i_\ell}; x_\ell)$, and $\sigma_1 \in S_{i_1}, \ldots, \sigma_\ell \in S_{i_\ell}$,*
$$(f \cdot (g_1\sigma_1, \ldots, g_\ell\sigma_\ell)) = (f \cdot (g_1, \ldots, g_\ell))\, \rho'(\sigma_1, \ldots, \sigma_\ell),$$
where $\rho'\colon S_{i_1} \times \cdots \times S_{i_\ell} \to S_{i_1 + \cdots + i_\ell}$ is the obvious homomorphism.

There is an obvious notion of a morphism from an S-operad O to an S-operad O': a function mapping each set $O(x_1, \ldots, x_\ell; x')$ to the corresponding set $O'(x_1, \ldots, x_\ell; x')$, preserving composition, identities, and the symmetric group actions. An important example is an 'algebra' of an S-operad:

DEFINITION 2. *For any S-operad O, an 'O-algebra' A consists of:*

1. *for any $x \in S$, a set $A(x)$.*
2. *for any $f \in O(x_1, \ldots, x_\ell; x')$, a function*
$$\alpha(f)\colon A(x_1) \times \cdots \times A(x_\ell) \to A(x')$$

such that:

(a) *whenever both sides make sense,*
$$\alpha(f \cdot (g_1, \ldots, g_\ell)) = \alpha(f)(\alpha(g_1) \times \cdots \times \alpha(g_\ell))$$
(b) *for any $x \in C$, $\alpha(1_x)$ acts as the identity on $A(x)$*

(c) *for any* $f \in O(x_1, \ldots, x_\ell, x')$ *and* $\sigma \in S_\ell$,

$$\alpha(f\sigma) = \alpha(f)\sigma,$$

where $\sigma \in S_\ell$ *acts on the function* $\alpha(f)$ *on the right by permuting its arguments.*

In what follows, by 'operad' we will mean an S-operad for some set S of types. We can think of such an operad as a simple sort of theory, and its algebras as models of this theory. Thus we can study operads either 'syntactically' or 'semantically'. To describe an operad syntactically, we list:

1. the set S of *types*,
2. the sets $O(x_1, \ldots, x_\ell; x')$ of *operations*,
3. the set of all *reduction laws* saying that some composite of operations (possibly with arguments permuted) equals some other operation.

This is like a presentation in terms of generators and relations, with the reduction laws playing the role of relations. On the other hand, to describe an operad semantically, we describe its algebras.

For example, the simplest operad is the 'initial untyped operad' I. Syntactically, this is the S-operad with:

1. only one type: $S = \{x\}$,
2. only one operation, the identity operation $1 \in O(x; x)$,
3. all possible reduction laws.

Semantically, I is the operad whose algebras are just sets.

Another important operad is the 'terminal untyped operad' T. This is the S-operad with

1. only one type: $S = \{x\}$,
2. exactly one operation of each arity,
3. all possible reduction laws.

The algebras of T are commutative monoids, with the ℓ-ary operation being ℓ-fold multiplication, or the unit element when $\ell = 0$, since nullary operations correspond to 'constants'.

5.2. Opetopes. The following fact is the key to defining the opetopes. Let O be an S-operad, and let $\mathrm{elt}(O)$ be the set of all operations of O.

THEOREM 3. *There is an* $\mathrm{elt}(O)$*-operad* O^+ *whose algebras are* S*-operads over* O, *i.e.,* S*-operads equipped with a homomorphism to* O.

We call O^+ the 'slice operad' of O. One can describe O^+ syntactically as follows:

1. The types of O^+ are the operations of O.
2. The operations of O^+ are the reduction laws of O.
3. The reduction laws of O^+ are the ways of combining reduction laws of O to give other reduction laws.

The 'level-shifting' going on here as we pass from O to O^+ is a way of systematizing the process of categorification.

A nice example of the slice operad construction is the operad I^+. The algebras of this operad turn out to be monoids. Another nice example is the operad T^+, whose algebras are untyped operads! It is also very interesting to iterate the slice operad construction. For example, the algebras of I^{++} work out to be exactly untyped planar operads. More generally, let O^{n+} be the operad formed by applying the slice operad construction n times to the operad O, or just O itself if $n = 0$.

DEFINITION 4. *An n-dimensional 'O-opetope' is a type of O^{n+}, or equivalently, if $n \geq 1$, an operation of $O^{(n-1)+}$.*

In particular, we define an n-dimensional 'opetope' to be an n-dimensional O-opetope for $O = I$, the initial untyped operad. The 0-dimensional opetopes are thus the types of I, but there is only one type, so there is only one 0-dimensional opetope, which we visualize as a point. The 1-dimensional opetopes are the types of I^+, or in other words, the operations of I. I has only one operation, the identity, so there is only one 1-dimensional opetope, which we visualize as an interval. The 2-dimensional opetopes are the types of I^{++}, or in other words, the operations of I^+, which are the reduction laws of I. These reduction laws all state that the identity operation composed with itself ℓ times equals itself. This leads to 2-dimensional opetopes with ℓ infaces and one outface. Actually there are $\ell!$ different 2-dimensional opetopes with ℓ infaces, since the permutation group S_ℓ acts freely on the set of ℓ-ary operations of I^+. We could keep track of these by labelling the infaces with some permutation of ℓ distinct symbols. A more systematic approach is to use 'metatree notation'. In this notation, any n-dimensional O-opetope is represented as a list of n labelled trees.

To see how this works, first consider the 2-dimensional O-opetopes, which are the operations of O^+. An operation of O^+ can be specified as in the figure below.

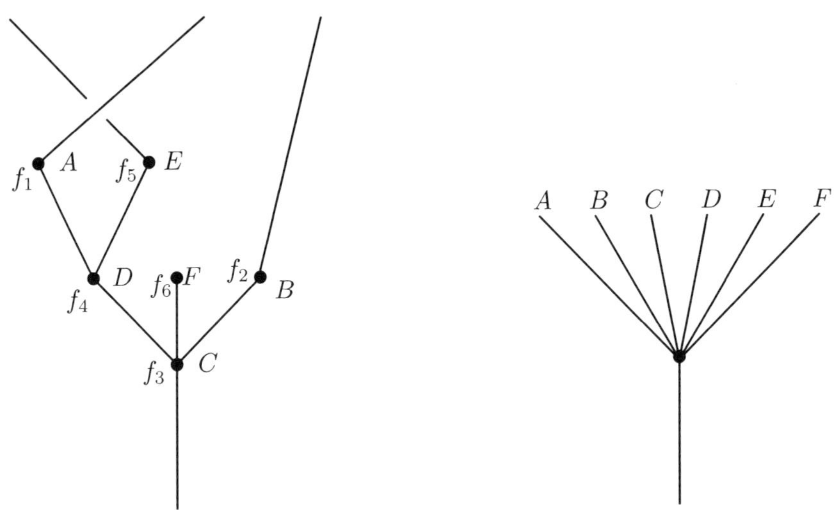

The first tree is an arbitrary 'O-tree'. This is a combed tree (i.e., planar except for a permutation of inputs at the top) with nodes labelled by operations of O. We require that a node labelled by a ℓ-ary operation have ℓ edges coming into it from above. Moreover, we require that it be possible to label every edge with an element of S in such a way that for any node labelled by an operation in $O(x_1, \ldots, x_\ell; x')$, the edges coming into that node from above are labelled by the types $x_1, \ldots, x_\ell$ in that order, while the edge coming out of it from below is labelled by the type x'. We also label each node of this first tree with a distinct symbol A, B, C, etc.. The second tree is planar and has only one node, with n edges coming into that node from above, labelled by the same symbols $A, B, C, \ldots$ in any order. These specify the order of the input types of the operation of O^+ we are describing.

More generally, for any $n > 1$ one can specify any n-dimensional O-opetope by means of an 'n-dimensional metatree'. Here is an example for $n = 3$:

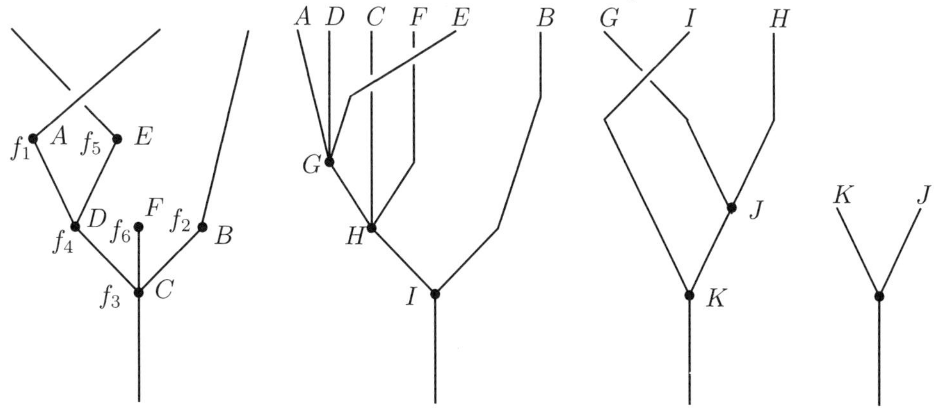

An n-dimensional metatree is a list of n labelled trees, the last of which is a planar tree with only one node, while the rest are combed trees. The first tree is an arbitrary O-tree. For $1 \leq i < n$, every node of the ith tree is labelled with a distinct symbol, and the same symbols also label all the edges at the very top of the $(i+1)$st tree, each symbol labelling exactly one edge. In addition, each edge of the $(i+1)$st tree must correspond to a subtree of the ith tree in such a way that:

1. The edge at the very top of the $(i+1)$st tree labelled by a given symbol corresponds to the subtree of the ith tree whose one and only node is labelled by the same symbol.
2. The edge of the $(i+1)$st tree coming out of a given node from below corresponds to the subtree that is the union of the subtrees corresponding to the edges coming into that node from above.
3. The edge at the very bottom of the $(i+1)$st tree corresponds to the whole ith tree.

Special care must be taken when the node of the last tree has no edges coming into it from above. This can only occur when all the previous trees are empty. This sort of metatree describes a nullary operation of $O^{(n-1)+}$ whose output type is an identity operation 1_x of $O^{(n-2)+}$. To specify which identity operation, we need to label the edge coming out of the node of the last tree from below with the operation 1_x.

6. n-Coherent operad algebras

An 'n-coherent O-algebra' is an n times categorified analog of an algebra of the operad O. In particular, when $O = I$, an n-coherent O-algebra is just an n-category, which is the n times categorified analog of a set. An n-coherent O-algebra is an 'O-opetopic set' with certain properties. We omit the precise definition of O-opetopic sets here. For our purposes, it should suffice to know that an O-opetopic set is very much like a simplicial set, but with O-opetopes replacing simplices, and no 'degeneracy maps', only 'face maps'. An O-opetopic set thus consists of (possibly empty) collections of 'cells' shaped like all the different O-opetopes, such that any inface or outface of a cell is again a cell.

If $j \geq 1$, we may schematically represent a j-dimensional cell x in an O-opetopic set as follows:

$$(a_1, \ldots, a_\ell) \xrightarrow{\quad x \quad} a'$$

Here $a_1, \ldots, a_\ell$ are the infaces of x and a' is the outface of x; all these are cells of one lower dimension. A configuration just like this, satisfying all the incidence relations satisfied by the boundary of a cell, but with x itself missing:

$$(a_1, \ldots, a_\ell) \xrightarrow{\quad ? \quad} a'$$

is called a 'frame'. A 'niche' is like a frame with the outface missing:

$$(a_1, \ldots, a_\ell) \xrightarrow{\quad ? \quad} ?$$

Similarly, a 'punctured niche' is like a frame with the outface and one inface missing:

$$(a_1, \ldots, a_{i-1}, ?, a_{i+1}, \ldots, a_\ell) \xrightarrow{\quad ? \quad} ?$$

If one of these configurations (frame, niche, or punctured niche) can be extended to an actual cell, the cell is called an 'occupant' of the configuration. Occupants of the same frame are called 'frame-competitors', while occupants of the same niche are called 'niche-competitors'.

Next we need the concept of a 'universal occupant' of a niche. Since the definition of this concept looks rather formidable at first, we first give a heuristic explanation. As already noted, the main use of universality is to define composites:

DEFINITION 5. *Given a universal occupant u of a j-dimensional niche:*

$$(a_1, \ldots, a_k) \xrightarrow{\quad u \quad} b$$

we call b a 'composite' of $(a_1, \ldots, a_k)$.

To understand universality more deeply, one must understand the role played by cells of different dimensions. In our framework an n-category usually has cells of arbitrarily high dimension, just like a Kan complex. For $j \leq n$ the j-dimensional cells play the role of j-morphisms, while for $j > n$ they play the role of 'equations', 'equations between equations', and so on. The definition of universality depends on n in a way that has the following effects. For $j \leq n$ there may be many universal occupants of a given j-dimensional niche, which is why we speak of 'a' composite rather than 'the' composite. There is at most one occupant of any given $(n+1)$-dimensional niche, which is automatically universal. Thus composites of n-cells are unique, and we may think of the universal occupant of an $(n+1)$-dimensional niche as an equation saying that the composite of the infaces equals the outface. For $j > n+1$ there is exactly one occupant of each j-dimensional frame, indicating that the composite of the equations corresponding to the infaces equals the equation corresponding to the outface.

The definition of universality essentially says that a j-dimensional niche-occupant is universal if all of its niche-competitors factor through it uniquely, *up to equivalence*. For $j \geq n+1$ this amounts to saying that each niche has a unique occupant, while for $j = n$ it means that each niche has an occupant through which all of its niche-competitors factor uniquely. Technically, the definition of universality says that composition with a universal niche-occupant set up a 'balanced punctured niche' of one higher dimension. One should think of a balanced punctured niche as defining an equivalence between occupants of its outface and occupants of its missing outface.

Now let us give the actual definition:

DEFINITION 6. *A j-dimensional niche-occupant:*

$$(c_1, \ldots, c_k) \xrightarrow{\quad u \quad} d$$

is said to be 'universal' if and only if $j > n$ and u is the only occupant of its niche, or $j \leq n$ and for any frame-competitor d' of d, the $(j+1)$-dimensional punctured niche

$$\begin{array}{c} ((c_1, \ldots, c_k) \xrightarrow{u} d,\ d \xrightarrow{?} d') \\ \Big\downarrow ? \\ (c_1, \ldots, c_k) \xrightarrow{?} d' \end{array}$$

and its mirror-image version

$$\begin{array}{c} (d \xrightarrow{?} d',\ (c_1, \ldots, c_k) \xrightarrow{u} d) \\ \Big\downarrow ? \\ (c_1, \ldots, c_k) \xrightarrow{?} d' \end{array}$$

are balanced.

Of course, now we need the definition of 'balanced'. The reader will note that while the definitions of 'universal' and 'balanced' call upon each other, there is no bad circularity.

DEFINITION 7. *An m-dimensional punctured niche:*

$$(a_1, \ldots, a_{i-1}, ?, a_{i+1}, \ldots, a_k) \xrightarrow{\quad ? \quad} ?$$

is said to be 'balanced' if and only if $m > n + 1$ or:

1. *any extension*

$$(a_1, \ldots, a_{i-1}, ?, a_{i+1}, \ldots, a_k) \xrightarrow{\quad ? \quad} b$$

extends further to:

$$(a_1, \ldots, a_{i-1}, a_i, a_{i+1}, \ldots, a_k) \xrightarrow{\quad u \quad} b$$

with u universal in its niche, and

2. *for any occupant*

$$(a_1, \ldots, a_{i-1}, a_i, a_{i+1}, \ldots, a_k) \xrightarrow{\quad u \quad} b$$

universal in its niche, and frame-competitor a'_i of a_i, the $(m+1)$-dimensional punctured niche

$$\begin{array}{c}(a'_i \xrightarrow{?} a_i,\ (a_1,\ldots,a_{i-1},a_i,a_{i+1},\ldots,a_k) \xrightarrow{u} b) \\ \Big\downarrow ? \\ (a_1,\ldots,a_{i-1},a'_i,a_{i+1},\ldots,a_k) \xrightarrow{?} b\end{array}$$

and its mirror-image version

$$\begin{array}{c}((a_1,\ldots,a_{i-1},a_i,a_{i+1},\ldots,a_k) \xrightarrow{u} b,\ a'_i \xrightarrow{?} a_i) \\ \Big\downarrow ? \\ (a_1,\ldots,a_{i-1},a'_i,a_{i+1},\ldots,a_k) \xrightarrow{?} b\end{array}$$

are balanced.

Note that the first numbered condition in the definition of 'balanced' definition generalizes the concept of an essentially surjective functor, while the second generalizes the concept of a fully faithful functor.

Finally, we define n-coherent O-algebras and various special cases:

DEFINITION 8. *An 'n-coherent O-algebra' is an O-opetopic set such that 1) every niche has a universal occupant, and 2) composites of universal cells are universal.*

DEFINITION 9. *An 'n-category' is an n-coherent I-algebra.*

DEFINITION 10. *A 'monoidal n-category' is an n-coherent I^+-algebra.*

DEFINITION 11. *A 'stable n-category' is an n-coherent T-algebra.*

One can show that any S-operad morphism $f\colon O \to O'$ allows one to turn an n-coherent O'-algebra A into an n-coherent O-algebra f^*A. Thus any stable n-category has an underlying monoidal n-category, and any monoidal n-category has an underlying n-category.

Given an n-category with 0-cells x and y, there is an $(n-1)$-category $\hom(x,y)$. One can also construct a stable $(n+1)$-category of all (small) n-categories, though the details of this construction have not yet been published. Using these facts, one can give rigorous formulations of many processes going between neighboring entries in Table 2: decategorification, discrete categorification, delooping, looping, forgetting monoidal structure, stabilization, and the generalized center construction. However, it remains to make precise and prove the stabilization hypothesis in this framework. Basically, one wishes to show that for $k \geq n+2$, the $(n+1)$-category of all stable n-categories is equivalent to the full sub-$(n+1)$-category of all n-categories having only one 0-cell and only one j-cell in each frame for $0 < j < k$.

7. Conclusions

In this paper we have discussed iterated categorifications and stabilizations of some of the very simplest algebraic structures: the natural numbers and the integers. However, one can also categorify many other concepts: vector spaces [**40**] and Hilbert spaces [**2**], group algebras [**21**], algebras of formal power series [**5, 37**] and other Hopf algebras [**20, 22**], sheaves [**15, 17**], and so on. Interesting results about these familiar structures typically have interesting categorified analogs. It is clear, therefore, that the set-based mathematics we know and love is just the tip of an immense iceberg of n-categorical, and ultimately ω-categorical, mathematics.

The prospect of exploring this huge body of new mathematics is both exhilarating and daunting. The basic philosophy is simple: *never mistake equivalence for equality.* The technical details, however, are not so simple — at least not yet. To proceed efficiently it is crucial that we gain a clearer understanding of the foundations before rushing ahead with complicated constructions.

Many basic questions remain open. For example, how significant is the fact that operads play a role both in the theory of E_k spaces and the definition of n-categories described above? Operads are very versatile, so this might at first seem to be a coincidence. However, there are deep relationships between operads, categorification, and the theory of algebraic structures satisfying laws 'up to coherent homotopy' [**60**]. In particular, Trimble [**64**] has pointed out an interesting connection. For many purposes it is best to think of E_k spaces as algebras, not of the little k-cubes operad, but of a closely related operad $F(k)$ discovered by Getzler and Jones [**33**]. The space $F(k)_\ell$ is the Fulton-MacPherson compactification of the configuration space of ℓ points in $\mathbb{R}^k$ modulo translations and dilations [**31**]. In particular, $F(1)_\ell$ is just $K_\ell \times S_\ell$, where K_ℓ is the $(\ell - 2)$-dimensional associahedron.

Since homotopy n-types of E_1 spaces correspond to monoidal n-groupoids, while n-coherent I^+-algebras are monoidal n-categories, one might expect a relationship between $F(1)$ and the I^+-opetopes. The associahedron K_ℓ has a cell decomposition having cells in one-to-one correspondence with planar trees with ℓ leaves for which all nodes have at least one edge coming in from above. It follows that the cells in the corresponding decomposition of $F(1)$ correspond to a certain class of 2-dimensional I^+-opetopes, or equivalently, 3-dimensional opetopes. Is there a deeper relation between opetopes and the associahedron? This might shed new light on the origin of coherence laws.

Acknowledgments. We thank Michael Batanin, Dan Christensen, Mark Hovey, André Joyal, Carlos Simpson, Charles Rezk, Ross Street, and Todd Trimble for helpful discussions and correspondence, and thank Ezra Getzler and Mikhail Kapranov both for useful correspondence and for organizing the conference at which this paper was presented.

References

[1] J. Adams, "Infinite Loop Spaces," Princeton U. Press, Princeton, 1978.

[2] J. Baez, *Higher-dimensional algebra II: 2-Hilbert spaces,* Adv. Math. **127** (1997), 125–189.

[3] J. Baez, *An introduction to n-categories,* 7th Conference on Category Theory and Computer Science, eds. E. Moggi and G. Rosolini, Lecture Notes in Computer Science **1290**, Springer Verlag, Berlin, 1997.

[4] J. Baez and J. Dolan, Higher-dimensional algebra and topological quantum field theory, *Jour. Math. Phys.* **36** (1995), 6073–6105.

[5] J. Baez and J. Dolan, Higher-dimensional algebra III: n-Categories and the algebra of opetopes, *Adv. Math.* **135** (1998), 145–206.
[6] J. Baez and L. Langford, 2-Tangles, *Lett. Math. Phys.*, 43 (1998), 187–197
[7] J. Baez and L. Langford, Higher-dimensional algebra IV: 2-Tangles, *Adv. Math.* **43** (1998), 187–197.
[8] J. Baez and M. Neuchl, Higher-dimensional algebra I: Braided monoidal 2-categories, *Adv. Math.* **121** (1996), 196–244.
[9] M. G. Barratt and S. Priddy, On the homology of non-connected monoids and their associated groups, *Comm. Math. Helv.* **47** (1972), 1–14.
[10] M. Batanin, Monoidal globular categories as a natural environment for the theory of weak n-categories, *Adv. Math.* **136** (1998), 39–103.
[11] H.-J. Baues, Homotopy types, in *Handbook of Algebraic Topology*, ed. I. M. James, Elsevier, New York, 1995.
[12] J. Bénabou, Introduction to bicategories, Lecture Notes in Mathematics **47**, Springer Verlag, Berlin, 1967, pp. 1–77.
[13] J. M. Boardman and R. M. Vogt, Homotopy-everything H-spaces, *Bull. Amer. Math. Soc.* **74** (1968), 1117–1122.
[14] J. M. Boardman and R. M. Vogt, *Homotopy invariant structures on topological spaces*, Lecture Notes in Mathematics **347**, Springer Verlag, Berlin, 1973.
[15] L. Breen, On the classification of 2-gerbes and 2-stacks, *Astérisque* **225** (1994), 1–160.
[16] R. Brown, From groups to groupoids: a brief survey, *Bull. London Math. Soc.* **19** (1987), 113–134.
[17] J.-L. Brylinski, *Loop Spaces, Characteristic Classes, and Geometric Quantization,* Birkhauser, Boston, 1993.
[18] R. Brown and C. Spencer, $\mathcal{G}$-groupoids, crossed modules, and the classifying space of a topological group, *Proc. Kon. Akad. v. Wet.* **79** (1976), 296–302.
[19] L. Crane, Clock and category: is quantum gravity algebraic?, *Jour. Math. Phys.* **36** (1995), 6180–6193.
[20] L. Crane and I. Frenkel, Four dimensional topological quantum field theory, Hopf categories, and the canonical bases, *Jour. Math. Phys.* **35** (1994), 5136–5154.
[21] L. Crane and D. Yetter, Examples of categorification, *Cahiers Topologie Gom. Diffrentielle Catg.* **39** (1998) , 3–25.
[22] L. Crane and D. Yetter, Deformations of (bi)tensor categories, preprint available as q-alg/9612011.
[23] S. E. Crans, On combinatorial models for higher dimensional homotopies, Ph.D. thesis, University of Utrecht, Utrecht, 1991.
[24] S. E. Crans, Generalized centers of braided and sylleptic monoidal 2-categories, *Adv. Math.* **136** (1998), 183–223.
[25] J. S. Carter, J. H. Rieger and M. Saito, A combinatorial description of knotted surfaces and their isotopies, *Adv. Math.* **127** (1997), 1–51.
[26] B. Day and R. Street, Monoidal bicategories and Hopf algebroids, *Adv. Math.* **129** (1997), 99–157.
[27] B. Eckmann and P. Hilton, Group-like structures in categories, *Math. Ann.* **145** (1962), 227–255.
[28] S. Eilenberg and G. M. Kelly, Closed categories, in *Proceedings of the Conference on Categorical Algebra,* eds. S. Eilenberg *et al*, Springer Verlag, New York, 1966.
[29] A. D. Elmendorf, I. Kriz, M. Mandell and J. P. May, Modern foundations for stable homotopy theory, in *Handbook of Algebraic Topology*, ed. I. M. James, Elsevier, New York, 1995.
[30] P. Freyd and D. Yetter, Braided compact monoidal categories with applications to low dimensional topology, *Adv. Math.* **77** (1989), 156–182.
[31] W. Fulton and R. MacPherson, A compactification of configuration spaces, *Ann. Math.* **139** (1994), 183–225.
[32] P. Gabriel and M. Zisman, *Calculus of Fractions and Homotopy Theory*, Ergebnisse der Math. und ihrer Grenzgebiete, vol. 35, Springer Verlag, Berlin, 1967.
[33] E. Getzler and J. Jones, Operads, homotopy algebra, and iterated integrals for double loop spaces, preprint available as hep-th/9403055.
[34] R. Gordon, A. J. Power, and R. Street, Coherence for tricategories, *Memoirs Amer. Math. Soc.* **117** (1995) Number 558.

[35] A. Grothendieck, Pursuing stacks, unpublished manuscript, 1983, distributed from UCNW, Bangor, United Kingdom.

[36] C. Hermida, M. Makkai, and J. Power, On weak higher dimensional categories, preprint available at http://hypatia.dcs.qmw.ac.uk/authors/M/MakkaiM/papers/multitopicsets/

[37] A. Joyal, Une théorie combinatoire des séries formelles, *Adv. Math.* **42** (1981), 1–82.

[38] A. Joyal and R. Street, Tortile Yang-Baxter operators in tensor categories, *J. Pure Appl. Alg.* **71** (1991), 43–51.

[39] A. Joyal and R. Street, Braided tensor categories, *Adv. Math.* **102** (1993), 20–78.

[40] M. Kapranov and V. Voevodsky, 2-Categories and Zamolodchikov tetrahedra equations, in *Proc. Symp. Pure Math.* **56** Part 2 (1994), AMS, Providence, pp. 177–260.
Braided monoidal 2-categories, 2-vector-spaces, and Zamolodchikov tetrahedra equations, Northwestern U. preprint.

[41] G. Kelly, Coherence theorems for lax algebras and distributive laws, Lecture Notes in Mathematics **420**, Springer Verlag, Berlin, 1974, pp. 281–375.

[42] G. Kelly and R. Street, Review of the elements of 2-categories, Lecture Notes in Mathematics **420**, Springer Verlag, Berlin, 1974, pp. 75–103.

[43] V. Kharlamov and V. Turaev, On the definition of the 2-category of 2-knots, *Amer. Math. Soc. Transl.* **174** (1996), 205–221.

[44] M. Laplaza, Coherence for distributivity, Lecture Notes in Mathematics **281**, Springer Verlag, Berlin, 1972, pp. 29–72.

[45] S. Mac Lane, Natural associativity and commutativity, *Rice U. Studies* **49** (1963), 28–46.

[46] S. Mac Lane, *Categories for the Working Mathematician*, Springer Verlag, New York, 1971.

[47] S. Mac Lane and J. Whitehead, On the 3-type of a complex, *Proc. Nat. Acad. Sci.* **36** (1950), 41–58.

[48] S. Majid, Representations, duals and doubles of monoidal categories, *Suppl. Rend. Circ. Mat. Palermo,* Series II **26** (1991), 197–206.

[49] J. P. May, *Simplicial Objects in Algebraic Topology*, Van Nostrand, Princeton, 1968.

[50] J. P. May, *The Geometry of Iterated Loop Spaces*, Lecture Notes in Mathematics **271**, Springer Verlag, Berlin, 1972.

[51] J. P. May, F. Quinn, N. Ray and J. Tornehave, *E_∞ Ring Spaces and E_∞ Ring Spectra*, Lecture Notes in Mathematics **577**, Springer Verlag, Berlin, 1977.

[52] A. J. Power, Why tricategories?, *Info. Comp.* **120** (1995), 251–262.

[53] D. Quillen, *Homotopical Algebra*, Lecture Notes in Mathematics **43**, Springer Verlag, Berlin, 1967.

[54] G.-C. Rota, *Gian-Carlo Rota on Combinatorics: Introductory Papers and Commentaries,* ed. J. P. S. Kung, Birkhauser, Boston, 1995.

[55] G. Segal, Configuration spaces and iterated loop-spaces, *Invent. Math.* **21** (1973), 213–222.

[56] M.-C. Shum, Tortile tensor categories, *Jour. Pure Appl. Alg.* **93** (1994), 57–110.

[57] C. Simpson, A closed model structure for n-categories, internal Hom, n-stacks and generalized Seifert-Van Kampen, preprint available as alg-geom/9704006.

[58] C. Simpson, Limits in n-categories, preprint available as alg-geom/9708010.

[59] J. D. Stasheff, Homotopy associativity of H-spaces I & II, *Trans. Amer. Math. Soc.* **108** (1963), 275–292, 293–312.

[60] J.-L. Loday, J. D. Stasheff, and A. A. Voronov, eds., *Operads: Proceedings of Renaissance Conferences,* A. M. S., Providence, Rhode Island, 1997.

[61] R. Street, The algebra of oriented simplexes, *Jour. Pure Appl. Alg.* **49** (1987), 283–335.

[62] Z. Tamsamani, Sur des notions de ∞-categorie et ∞-groupoide non-strictes via des ensembles multi-simpliciaux, preprint available as alg-geom/9512006.

[63] Z. Tamsamani, Equivalence de la théorie homotopique des n-groupoides et celle des espaces topologiques n-tronqués, preprint available as alg-geom/9607010.

[64] T. Trimble, personal communication.

[65] V. Turaev, Operator invariants of tangles, and R-matrices, *Math. USSR Izvestia* **35** (1990), 411–444.

[66] D. Yetter, Markov algebras, in *Braids*, eds. J. S. Birman and A. Libgober, Contemporary Mathematics **78**, A. M. S., Providence, Rhode Island, 1988, 705–730.

DEPARTMENT OF MATHEMATICS, UNIVERSITY OF CALIFORNIA, RIVERSIDE, CALIFORNIA 92521
E-mail address: baez@math.ucr.edu

SANGA RESEARCH CORPORATION, 2015 RUE PEEL, MONTREAL, QUEBEC H3A 1T8, CANADA
E-mail address: jdolan@sangacorp.com

Contemporary Mathematics
Volume **230**, 1998

Computads for finitary monads on globular sets

M. A. Batanin

ABSTRACT. A finitary monad A on the category of globular sets provides basic algebraic operations from which more involved 'pasting' operations can be derived. To makes this rigorous, we define A-computads and construct a monad on the category of A-computads whose algebras are A-algebras; an action of the new monad encapsulates the pasting operations. When A is the monad whose algebras are n-categories, an A-computad is an n-computad in the sense of R. Street. When A is associated to a higher operad (in the sense of the author), we obtain pasting in weak n-categories. This is intended as a first step towards proving the equivalence of the various definitions of weak n-category now in the literature.

Introduction

This work arose as a reflection on the foundation of higher dimensional category theory. One of the main ingredients of any proposed definition of weak n-category is the shape of diagrams (pasting scheme) we accept to be composable. In a globular approach [**3**] each k-cell has a source and target $(k-1)$-cell. In the opetopic approach of Baez and Dolan [**1**] and the multitopic approach of Hermida, Makkai and Power [**7**] each k-cell has a unique $(k-1)$-cell as target and a whole $(k-1)$-dimensional pasting diagram as source. In the theory of strict n-categories both source and target may be a general pasting diagram [**9, 14, 15**].

The globular approach being the simplest one seems too restrictive to describe the combinatorics of higher dimensional compositions. Yet, we argue that this is a false impression. Moreover, we prove that this approach is a basic one from which the other type of composable diagrams may be derived. One theorem proved here asserts that the category of algebras of a finitary monad on the category of n-globular sets is **equivalent** to the category of algebras of an appropriate monad on the special category (of computads) constructed from the data of the original monad. In the case of the monad derived from the universal contractible operad [**3**] this result may be interpreted as the equivalence of the definitions of weak n-categories (in the sense of [**3**]) based on the 'globular' and general pasting diagrams. It may be also considered as the first step toward the proof of equivalence of the different definitions of weak n-category.

1991 *Mathematics Subject Classification.* Primary 18C20, 18D05.
I am grateful to Australian Research Council for financial support (Large Grant # A69700282).

We also develop a general theory of computads and investigate some properties of the category of generalized computads. It turned out, that in a good situation this category is a topos (and even a presheaf topos under some not very restrictive conditions, the property firstly observed by S. Schanuel and reproved in [**4**] for 2-computads in the sense of Street).

I am grateful to C. Berger, R. Buckland, S. Crans, M. McIntyre, R. Street for valuable discussions. To R. Street my special thanks are due for his constant support at every stage of work and aid in preparation of this paper for publishing. I am also grateful to my referee for helpful suggestions for improving the original version of the paper.

1. Preliminary discussion

Computads were defined by R. Street [**12, 14**] for the purposes of the presentation of (strict) n-categories. The definition is inductive. Let us give it here.

A 0-computad is a set, a free 0-category on it is this set itself.

A 1-computad is a directed graph, a free 1-category on it is the usual category of directed paths in that graph.

Suppose we know what an $(n-1)$-computad $\mathcal{C}$ is and have a construction of a free (strict) $(n-1)$-category $\mathcal{F}_{n-1}\mathcal{C}$ on it. Then an n-computad consists of a set C_n of n-cells, an $(n-1)$-computad $\mathcal{C}$ and two functions (source and target)

$$s_{n-1}, t_{n-1} : C_n \longrightarrow (\mathcal{F}_{n-1}\mathcal{C})_{n-1}$$

where $(\mathcal{F}_{n-1}\mathcal{C})_{n-1}$ is the set of $(n-1)$-cells in $\mathcal{F}_{n-1}\mathcal{C}$. These functions must satisfy the equations

$$\text{(1.1)} \qquad s_{n-2} \cdot s_{n-1} = s_{n-2} \cdot t_{n-1} \ , \ t_{n-2} \cdot s_{n-1} = t_{n-2} \cdot t_{n-1},$$

where s_{n-2}, t_{n-2} are source and target functions in $\mathcal{F}_{n-1}\mathcal{C}$.

The n-computads form a category, where a morphism of n-computads consists of a morphism between corresponding $(n-1)$-computads and a map between the sets of n-cells which agree one to another via the source and target functions.

Computads turned out to be an excellent device for working with presentations of n-categories and pasting operations (see [**14**] for many examples), especially in the dimension 2 where R. Street proved that 2-categories may be characterised in terms of an abstract pasting operation on an appropriate 2-computad. More precisely, the forgetful functor $\mathcal{W}_2$ from 2-computads to 2-categories is monadic [**12**].

The 2-computads appear also in the calculus of string diagrams [**8**]. To explain this connection look at a typical 2-cell in a free 2-category generated by a terminal 2-computad:

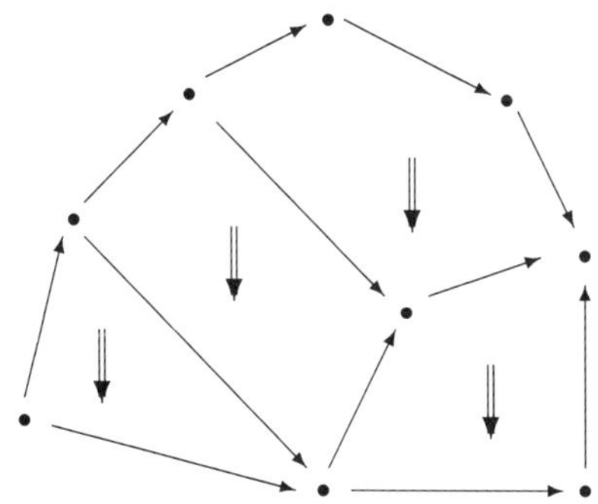

If we take a planar dual of this diagram we obtain a topological graph that is usually called string diagram [**14**]. A deep theorem of A. Joyal and R. Street [**8**] asserts that the free 2-category on the terminal 2-computad is isomorphic to the 2-category of deformation classes of progressive plane graphs.

However, if we try to go on one dimension up and develop the calculus of surface diagrams in 3-dimensional space we find that 3-computads are not sufficient. The right notion for this is Gray-computad in the sense of [**11**]. Again we have a theorem that a free Gray-category on a terminal Gray-computad is isomorphic to the Gray-category of deformation classes of surface diagrams [**11**].

This suggests that higher dimensional analogues of Gray-computad might be very important as they, probably, connect higher category theory to the geometry of real space.

Furthemore, the notion of computad appears in connection with the appropriate notion of higher-dimensional category. According to [**3**] these categories are algebras of an appropriate operad (in the sense of [**3**]) on the category of globular sets. By a globular set we mean a sequence of sets

$$X_0, X_1, \dots, X_r, \dots$$

together with source and target maps

$$s_{n-1}, t_{n-1} : X_n \longrightarrow X_{n-1}$$

satisfying the equation (1.1). Every operad on globular sets generates a finitary monad whose algebras are algebras of the operad. In the case of the terminal operad this monad assigns to a globular set X the underlying globular set of the free ω-category generated by X (see [**3**] or [**16**] for the description of this monad in terms of the combinatorics of trees). Thus we expect more generally, that every finitary monad on globular sets generates its own notion of computad. In the case of terminal operad we should get Street's n-computads. We prove this in the next sections.

2. Definition of computad

Let $\mathcal{G}l_n$ be the subcategory of *Set* which has as objects the sets $\{0, k\}$, $0 \le k \le n$ and only the constant function $\{0, k\} \to \{0, l\}$, $k > l$, as nonidentity morphisms. The category $Glob_n$ of n-globular sets (the category *Glob* of globular sets for $n = \infty$) is the category of functors

$$\mathcal{G}l_n \to Set.$$

For an n-globular set C we write C_k for $C\{0,k\}$ and s_l, t_l for the image under C of the functions $\{0,k\} \to \{0,l\}$, $k > l$, constant at $0, l$, respectively.

For every $k \leq n$, we have a functor

$$L_k : Glob_k \to Glob_n$$

which assigns to every k-globular set X a globular set $L_k X$ which coincides with X up to dimension k and has empty sets in all dimensions greater than k. This functor has a right adjoint tr_k which assigns the k-truncation to each globular set. The counit of this adjunction will be denoted by $\wp_k$.

Let $A_n = (A_n, \mu_n, \epsilon_n)$ be a finitary monad on $Glob_n$ (recall that monad is called finitary if its functor part preserves filtered colimits [**10**]). For every $k \leq n$, we have a monad $A_k = (A_k, \mu_k, \epsilon_k)$ on $Glob_k$ with underlying functor $\mathrm{tr}_k A_n L_k$, multiplication

$$\mathrm{tr}_k A_n L_k \,\mathrm{tr}_k A_n L_k \xrightarrow{\wp_k} \mathrm{tr}_k A_n^2 L_k \xrightarrow{\mu_n} \mathrm{tr}_k A_n L_k$$

and unit

$$I \xrightarrow{id} \mathrm{tr}_k L_k \xrightarrow{\epsilon k} \mathrm{tr}_k A_n L_k.$$

We denote by Alg_k the category of algebras of A_k, by $W_k : Alg_k \to Glob_k$ the corresponding forgetful functor and by $F_k : Glob_k \to Alg_k$ the left adjoint of W_k. The functor tr_k obviously induces a functor

$$Alg_n \longrightarrow Alg_k$$

which will be denoted by tr_k as well.

Let us give the following inductive definition:

The category $Comp_0$ of A_0-computads is $Glob_0$. The functors

$$\mathcal{W}_0 = W_0 : Alg_0 \to Comp_0$$

$$\mathcal{F}_0 = F_0 : Comp_0 \to Alg_0$$

are forgetful and free A_0-algebra functors, respectively.

Let us suppose now that the category $Comp_{n-1}$ of A_{n-1}-computads is already defined together with two functors:

$$\mathcal{W}_{n-1} : Alg_{n-1} \to Comp_{n-1}$$

$$\mathcal{F}_{n-1} : Comp_{n-1} \to Alg_{n-1}$$

such that $\mathcal{F}_{n-1}$ is left adjoint to $\mathcal{W}_{n-1}$.

DEFINITION 2.1. An A_n-computad $\mathcal{C}$ is a triple $(C, \phi, \mathcal{C}')$ consisting of an n-globular set C, an A_{n-1}-computad $\mathcal{C}'$ and an isomorphism ϕ in $Glob_{n-1}$

$$\phi : W_{n-1}(\mathcal{F}_{n-1}\mathcal{C}') \to \mathrm{tr}_{n-1} C.$$

A morphism

$$F : (C, \phi, \mathcal{C}') \to (D, \psi, \mathcal{D}')$$

of A_n-computads is a pair (f, Φ) where $f : C \to C'$ is a morphism of n-globular sets and $\Phi : \mathcal{C}' \to \mathcal{D}'$ is a morphism of A_{n-1}-computads such that the following diagram commutes

$$\begin{array}{ccc} W_{n-1}(\mathcal{F}_{n-1}\mathcal{C}') & \xrightarrow{W_{n-1}(\mathcal{F}_{n-1}\Phi)} & W_{n-1}(\mathcal{F}_{n-1}\mathcal{D}') \\ \phi \downarrow & & \downarrow \psi \\ \mathrm{tr}_{n-1} C & \xrightarrow{\mathrm{tr}_{n-1} f} & \mathrm{tr}_{n-1} D \end{array}$$

The most important case for us are the monads coming from higher operads in the sense of [**3**]. We refer the reader to [**3, 16**] for the definition. The brief explanation is given in the example 2 of section 6 of this paper.

DEFINITION 2.2. For an n-operad $\mathcal{A}_n$ in $Span_n$, the category of $\mathcal{A}_n$-computads is the category of A_n-computads, where A_n is the monad on $Glob_n$ generated by $\mathcal{A}_n$.

3. Existence

To give sense to this definition we have to construct an adjoint pair of functors

$$\mathcal{F}_n \dashv \mathcal{W}_n.$$

Let G be an object of Alg_n.The counit of the adjunction $\mathcal{F}_{n-1} \dashv \mathcal{W}_{n-1}$ gives a morphism

$$r_{n-1} : \mathcal{F}_{n-1}\mathcal{W}_{n-1}\,\mathrm{tr}_{n-1} G \to \mathrm{tr}_{n-1} G.$$

Define an n-globular set $\mathcal{G}$ in the following way. The $(n-1)$-skeleton of $\mathcal{G}$ coincides with $W_{n-1}\mathcal{F}_{n-1}\mathcal{W}_{n-1}\,\mathrm{tr}_{n-1} G$ and

$$\mathcal{G}_n = \{(\xi, a, \eta) \in \mathcal{G}_{n-1} \times G_n \times \mathcal{G}_{n-1} \mid s_{n-2}\xi = s_{n-2}\eta,\ t_{n-2}\xi = t_{n-2}\eta,$$
$$s_{n-1}a = r_{n-1}(\xi),\ t_{n-1}a = r_{n-1}(\eta)\}.$$

Define

$$s_{n-1}(\xi, a, \eta) = \xi\ ,\ t_{n-1}(\xi, a, \eta) = \eta.$$

Then put

$$\mathcal{W}_n G = (\mathcal{G}, id, \mathcal{W}_{n-1}\,\mathrm{tr}_{n-1} G).$$

Now our goal is to construct a left adjoint $\mathcal{F}_n$ for $\mathcal{W}_n$.

For an A_n-computad $\mathcal{C} = (C, \phi, \mathcal{C}')$, define

$$V_n(\mathcal{C}) = C$$

and $V_0 = id$ for $n = 0$. Define also the truncation functor

$$\mathrm{tr}_{n-1} : Comp_n \longrightarrow Comp_{n-1}$$

by

$$\mathrm{tr}_{n-1}\mathcal{C} = \mathcal{C}'.$$

Construct firstly a natural transformation

$$\Theta_n : V_n\mathcal{W}_n \to W_n.$$

We define $\Theta_n(G)$ to be the morphism of n-globular sets which coincides with

$$W_{n-1}r_{n-1} : W_{n-1}\mathcal{F}_{n-1}\mathcal{W}_{n-1}\,\mathrm{tr}_{n-1} G \to W_{n-1}\,\mathrm{tr}_{n-1} G$$

up to dimension $n-1$ and has

$$\Theta_n(\xi, a, \eta) = a$$

in the dimension n.

From the adjunctions $\mathcal{F}_{n-1} \dashv \mathcal{W}_{n-1}$, $F_{n-1} \dashv W_{n-1}$ we have the following mates

$$\frac{\dfrac{\Theta_{n-1} : V_{n-1}\mathcal{W}_{n-1} \to W_{n-1}}{\Omega_{n-1} : V_{n-1} \to W_{n-1}\mathcal{F}_{n-1}}}{\Upsilon_{n-1} : F_{n-1}V_{n-1} \to \mathcal{F}_{n-1}.}$$

We also have an A_{n-1}-algebra morphism

$$\beta : A_{n-1}\,\mathrm{tr}_{n-1}\,C = \mathrm{tr}_{n-1}\,A_n L_{n-1}\,\mathrm{tr}_{n-1}\,C \overset{\mathrm{tr}_{n-1}A_n\wp_{n-1}}{\longrightarrow} \mathrm{tr}_{n-1}\,A_n C.$$

Consider the following pair of composites

$$A_{n-1}V_{n-1}\,\mathrm{tr}_{n-1} \underset{A_{n-1}\epsilon_{n-1}}{\overset{\epsilon_{n-1}}{\rightrightarrows}} A^2_{n-1}V_{n-1}\,\mathrm{tr}_{n-1} \xrightarrow{\ \Psi\ } \mathrm{tr}_{n-1}\,A_n V_n$$

where Ψ is the composite

$$\Psi = \beta \cdot A_{n-1}\phi \cdot A_{n-1}W_{n-1}\Upsilon_{n-1}$$

This pair is mated with a pair

$$L_{n-1}A_{n-1}V_{n-1}\,\mathrm{tr}_{n-1} \underset{d_1}{\overset{d_0}{\rightrightarrows}} A_n V_n \tag{3.1}$$

which in its turn is mated to

$$F_n L_{n-1}A_{n-1}V_{n-1}\,\mathrm{tr}_{n-1} \underset{\partial_1}{\overset{\partial_0}{\rightrightarrows}} F_n V_n$$

We define

$$\Upsilon_n : F_n V_n \to \mathcal{F}_n$$

to be the coequalizer of (∂_0, ∂_1) (which exists due to finitary assumption [**10**].)

PROPOSITION 3.1. *The functor $\mathcal{F}_n$ is left adjoint to $\mathcal{W}_n$.*

PROOF. Let G be an A_n-algebra and

$$(f, \Phi) : \mathcal{C} \to \mathcal{W}_n G$$

be an A_n-computad morphism. So there exists a unique A_n-algebra morphism

$$\bar{f} : F_n V_n \mathcal{C} \to G$$

fitting commutatively into the diagram

$$\begin{array}{ccc}
V_n\mathcal{C} & \xrightarrow{\ \epsilon_n\ } & A_n V_n \mathcal{C} \\
{\scriptstyle f}\downarrow & & \downarrow{\scriptstyle W_n\bar{f}} \\
V_n \mathcal{W}_n G & \xrightarrow{\ \Theta_n\ } & W_n G
\end{array}$$

We now have to prove that

$$W_n\bar{f} \cdot d_0 = W_n\bar{f} \cdot d_1. \tag{3.2}$$

It is obvious that we need to prove (3.2) only for $(n-1)$-truncation of these morphisms. That is we need to prove that $\mathrm{tr}_{n-1} W_n\bar{f}$ coequalizes the diagram (3.1) after truncation.

We have the following diagram

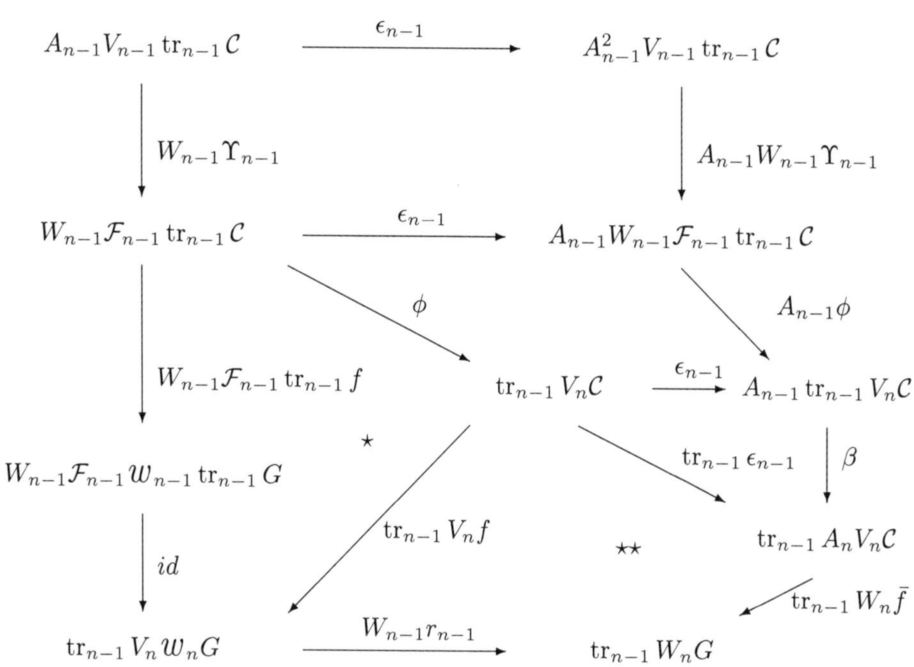

In this diagram the square $(\star)$ commutes by definition of morphism of computads, the square $(\star\star)$ commutes by definition of $\bar{f}$. The others commute by naturality. Hence, the whole diagram commutes.

We have thus proved the composition $\mathrm{tr}_{n-1} W_n\bar{f} \cdot \beta \cdot A_{n-1}\phi \cdot A_{n-1}W_{n-1}\Upsilon_{n-1}$ is an A_{n-1}-algebras morphism. So

$$\mathrm{tr}_{n-1} W_n\bar{f} \cdot \beta \cdot A_{n-1}\phi \cdot A_{n-1}W_{n-1}\Upsilon_{n-1} = \mathrm{tr}_{n-1} W_n\bar{f} \cdot \mathrm{tr}_{n-1} d_0 \cdot \mu_{n-1}.$$

And we obtain

$$\mathrm{tr}_{n-1} W_n\bar{f} \cdot \mathrm{tr}_{n-1} d_1 = \mathrm{tr}_{n-1} W_n\bar{f} \cdot \beta \cdot A_{n-1}\phi \cdot A_{n-1}\,\mathrm{tr}_{n-1}\,\rho_{n-1} \cdot A_{n-1}\epsilon_{n-1} =$$
$$\mathrm{tr}_{n-1} W_n\bar{f} \cdot \mathrm{tr}_{n-1} d_0 \cdot \mu_{n-1} \cdot A_{n-1}\epsilon_{n-1} = \mathrm{tr}_{n-1} W_n\bar{f} \cdot \mathrm{tr}_{n-1} d_0.$$

Hence, equation (3.2) is proved. So the morphism $\bar{f}$ can be extended uniquely to a morphism of A_n-algebras

$$\mathcal{F}_n\mathcal{C} \to G.$$

It is not hard to see that this construction generates a natural bijection

$$\mathit{Comp}_n(\mathcal{C}, \mathcal{W}_nG) \simeq \mathit{Alg}_n(\mathcal{F}_n\mathcal{C}, G)$$

as required. □

DEFINITION 3.1. We say that A_n is truncable in dimension k if

$$\beta : A_k \operatorname{tr}_k \longrightarrow \operatorname{tr}_k A_{k+1}$$

is an invertible natural transformation. We say that A_n is truncable if it is truncable in every dimension.

For a truncable A_n we can give the following alternative construction of the left adjoint $\mathcal{F}_n$.

Let $\mathcal{C}$ be an A_n-computad. Let

$$M_{-1}\mathcal{C} = L_{n-1}W_{n-1}\mathcal{F}_{n-1}\operatorname{tr}_{n-1}\mathcal{C}\ ,\ \ M_0\mathcal{C} = V_n\mathcal{C}$$

and

$$\phi_0 = \phi : \operatorname{tr}_{n-1} M_{-1}\mathcal{C} \to \operatorname{tr}_{n-1} M_0\mathcal{C}.$$

Notice that ϕ_0 transports the A_{n-1}-algebra structure from $\operatorname{tr}_{n-1} M_{-1}\mathcal{C}$ to $\operatorname{tr}_{n-1} M_0\mathcal{C}$.

Suppose that a globular set $M_r\mathcal{C}$, together with an A_{n-1}-algebra structure on $\operatorname{tr}_{n-1} M_r\mathcal{C}$ and an A_{n-1}-algebras isomorphism

$$\phi_r : \operatorname{tr}_{n-1} M_{r-1}\mathcal{C} \to \operatorname{tr}_{n-1} M_r\mathcal{C},$$

are already constructed. Then define $M_{r+1}\mathcal{C}$ to be the pushout of the following diagram

$$\begin{array}{ccc} L_{n-1}A_{n-1}\operatorname{tr}_{n-1} M_r\mathcal{C} & \xrightarrow{\gamma_r} & A_nM_r\mathcal{C} \\ {\scriptstyle L_{n-1}k}\downarrow & & \downarrow{\scriptstyle p_r} \\ L_{n-1}\operatorname{tr}_{n-1} M_r\mathcal{C} & \xrightarrow{\alpha_{r+1}} & M_{r+1}\mathcal{C} \end{array}$$

where k is the structure morphism for the A_{n-1}-algebra $\operatorname{tr}_{n-1} M_r\mathcal{C}$ and γ_r is the composite $\wp_{n-1} \cdot L_{n-1}\beta$. Define also

$$\phi_{r+1} = \operatorname{tr}_{n-1} \alpha_{r+1},$$

which is invertible as β is invertible and, hence, determines an A_{n-1}-algebra structure on $\operatorname{tr}_{n-1} M_{r+1}$.

Then we have the following sequence of morphisms

$$M_0\mathcal{C} \xrightarrow{\epsilon_n} A_nM_0\mathcal{C} \xrightarrow{p_1} M_1\mathcal{C} \xrightarrow{\epsilon_n} A_nM_2\mathcal{C} \xrightarrow{p_2} \dots .$$

As A_n is finitary, the colimit of this sequence has a canonical A_n-algebra structure induced by the following commutative diagrams for every r.

$$\begin{array}{ccccc} A_n^2M_r & \xrightarrow{A_np_{r+1}} & A_nM_{r+1} & \xrightarrow{A_n\epsilon_n} & A_n^2M_{r+1} \\ {\scriptstyle \mu_n}\downarrow & & & & \downarrow{\scriptstyle \mu_n} \\ A_nM_r & \xrightarrow{p_{r+1}} & M_{r+1} & \xrightarrow{\epsilon_n} & A_nM_{r+1} \end{array} \tag{3.3}$$

PROPOSITION 3.2. *The* $\operatorname{colim}_r M_r\mathcal{C}$ *with* A_n*-algebra structure as above is canonically isomorphic to* $\mathcal{F}_n\mathcal{C}$.

PROOF. Let G be an A_n-algebra and let

$$f : \mathcal{C} \longrightarrow \mathcal{W}_n G$$

be a morphism of A_n-computads. Define H_0 to be the following composite

$$M_0\mathcal{C} = V_n\mathcal{C} \xrightarrow{V_n f} V_n \mathcal{W}_n G \xrightarrow{\Theta_n} W_n G$$

We have the following commutative diagram after truncation

$$\begin{array}{ccccc}
\operatorname{tr}_{n-1} M_0 & \xrightarrow{\operatorname{tr}_{n-1} f} & \operatorname{tr}_{n-1} V_n \mathcal{W}_n G & \xrightarrow{\operatorname{tr}_{n-1} \Theta_n} & \operatorname{tr}_{n-1} W_n G \\
\Big\downarrow \phi & & \searrow id & & \nearrow W_{n-1} r_{n-1} \\
W_{n-1}\mathcal{F}_{n-1} \operatorname{tr}_{n-1} \mathcal{C} & \xrightarrow{W_{n-1}\mathcal{F}_{n-1} \operatorname{tr}_{n-1} f} & W_{n-1}\mathcal{F}_{n-1}\mathcal{W}_{n-1} \operatorname{tr}_{n-1} G & &
\end{array}$$

Hence, $\operatorname{tr}_{n-1} H_0$ is an A_{n-1}-algebras morphism.

Suppose we have constructed a morphism

$$H_r : M_r\mathcal{C} \longrightarrow W_n G$$

such that $\operatorname{tr}_{n-1} H_r$ is an A_{n-1}-algebras morphism. Then we can lift H_r to an A_n-algebra morphism

$$\bar{H}_r : A_n M_r \mathcal{C} \longrightarrow W_n G$$

As $\operatorname{tr}_{n-1} H_r$ is a morphism of A_{n-1}-algebras, we have that $\operatorname{tr}_{n-1} \bar{H}_r \cdot \beta$ is equal to the following composite

$$A_{n-1} \operatorname{tr}_{n-1} M_r\mathcal{C} \xrightarrow{k} \operatorname{tr}_{n-1} M_r \xrightarrow{\operatorname{tr}_{n-1} H_r} \operatorname{tr}_{n-1} W_n G.$$

We get, therefore, the morphism

$$L_{n-1} \operatorname{tr}_{n-1} M_r\mathcal{C} \longrightarrow W_n G,$$

which together with $\bar{H}_r$ determine the morphism

$$H_{r+1} : M_{r+1}\mathcal{C} \longrightarrow W_n G.$$

It is not hard to check that $\operatorname{tr}_{n-1} H_{r+1}$ is an A_{n-1}-algebra morphism. Hence, the induction works and we have a morphism

$$H : \operatorname{colim}_r M_r\mathcal{C} \longrightarrow W_n G.$$

Moreover, H is an A_n-algebra morphism as it is also the colimit of $\bar{H}_r$. It is now obvious how to recover a computad morphism f for a given A_n-algebra morphism H.

We have thus proved that $\operatorname{colim}_r M_r$ is left adjoint to $\mathcal{W}_n$ and, hence, that is canonically isomorphic to $\mathcal{F}_n$. □

COROLLARY 3.1. *For a truncable finitary monad* A_n *there exists a natural isomorphism*

$$\mathcal{F}_k \operatorname{tr}_k \longrightarrow \operatorname{tr}_k \mathcal{F}_{k+1}$$

PROOF. The sequence of isomorphisms ϕ_r gives the desired result. □

4. Some properties of computads

The results of this section are based on the theory developed in [**4**]. Recall the necessary definitions.

DEFINITION 4.1. Let $\phi : F \longrightarrow G$ be a natural transformation between two functors. We call it cartesian provided every naturality square

$$\begin{array}{ccc} F(a) & \xrightarrow{\phi} & G(a) \\ {\scriptstyle F(f)}\downarrow & & \downarrow{\scriptstyle G(f)} \\ F(b) & \xrightarrow{\phi} & G(b) \end{array}$$

is a pullback.

It is obvious that the composition of cartesian transformations is cartesian and if the composition $\phi \cdot \psi$ is cartesian and ϕ is cartesian then ψ is cartesian as well.

The following lemma is just a slight generalization of the fact, that a pullback of a cartesian natural transformation is cartesian and may be proved by the same method.

LEMMA 4.1. *Let P be a small category, F, G two functors from a category A to the presheaf category Set^P, and $f : F \to G$ a cartesian natural transformation. Let B be a functor from A to Set and $\phi_p(a)$ be a cone from $B(a)$ to $G(a)$ which is natural in a. Then we have a functor L to the category of sets which, for a fixed object a, is the limit of the diagram*

$$\begin{array}{ccc} L(a) & \xrightarrow{f^*} & B(a) \\ {\scriptstyle \psi}\downarrow & & \downarrow{\scriptstyle \phi} \\ F(a) & \xrightarrow{f_a} & G(a) \end{array}$$

We have also a natural transformation

$$f^* : L \longrightarrow B.$$

Then we claim that f^* is cartesian.

We shall call a *wide pullback* any limit indexed by a category which is a poset with a terminal object, in which all maximal chains have length 2.

DEFINITION 4.2. A monad $A = (A, \mu, \epsilon)$ on a category C preserves pullbacks (wide pullbacks) provided its functor part preserves them. It is called cartesian if it preserves pullbacks and ϵ and μ are cartesian natural transformations.

THEOREM 4.1. *Let A_n be a truncable finitary monad on $Glob_n$.*

(i) If A_n preserves (finite) pullbacks then $\mathcal{F}_n$ preserves them and $Comp_n$ is an elementary topos,

(ii) If A_n preserves wide pullbacks and ϵ_n is a cartesian natural transformation then $\mathcal{F}_n$ preserves them and $Comp_n$ is a presheaf topos.

PROOF. The proof generalizes the example 3.6 from [**4**]. We use induction on n. If $n = 0$ the statement is true by definition.

Suppose we know that the category $Comp_{n-1}$ is a topos (a presheaf topos) and, in addition, $\mathcal{F}_{n-1}$ preserves pullbacks (wide pullbacks). Consider the functor

$$T_{n-1} : Comp_{n-1} \longrightarrow Set ,$$

which assigns to a computad $\mathcal{C}$ the set of the parallel pairs of $(n-1)$-cells from $W_{n-1}\mathcal{F}_{n-1}\mathcal{C}$. Then we have the equivalence of categories

$$Comp_n \sim Set \downarrow T_{n-1}.$$

Now, prove that T_{n-1} preserves pullbacks (for (i)) or wide pullbacks (for (ii)). Notice that T_{n-1} is isomorphic to the following composite

$$Comp_{n-1} \stackrel{\mathcal{F}_{n-1}}{\longrightarrow} Alg_{n-1} \stackrel{Alg_{n-1}(F_{n-1}S^{n-1},-)}{\longrightarrow} Set$$

where S^{n-1} is an $n-1$-globular set, which has two elements $-$ and $+$ in every dimension and

$$s(-) = s(+) = - \ , \ t(-) = t(+) = +.$$

By the inductive assumption, $\mathcal{F}_{n-1}$ preserves pullbacks (wide pullbacks), so T_{n-1} does. According to the results of [**4**] it is sufficient for $Set \downarrow T_{n-1}$ to be a topos (preasheaf topos).

It remains to prove that $\mathcal{F}_n$ preserves pullbacks (wide pullbacks). Let us prove, firstly, that the functors M_r from proposition 3.2 preserve pullbacks (wide pullbacks in the case (ii)). Use induction on r. For $r = 0$ we have to prove that V_n preserves pullbacks (wide pullbacks). But that is obvious as we know that $\mathcal{F}_{n-1}$ preserves pullbacks (wide pullbacks) by our inductive assumption.

Let $\mathcal{C} = \lim_\alpha \mathcal{C}_\alpha$ be a pullback (wide pullback) in $Comp_n$. Then

$$\begin{aligned}\mathrm{tr}_{n-1} \lim_\alpha(M_{r+1}\mathcal{C}_\alpha) &\simeq \lim_\alpha(\mathrm{tr}_{n-1} M_{r+1}\mathcal{C}_\alpha) \simeq \lim_\alpha(W_{n-1}\mathcal{F}_{n-1}\,\mathrm{tr}_{n-1}\mathcal{C}_\alpha) \\ &\simeq W_{n-1}\mathcal{F}_{n-1} \lim_\alpha(\mathrm{tr}_{n-1}\mathcal{C}_\alpha) \simeq \mathrm{tr}_{n-1} M_{r+1} \lim_\alpha(\mathcal{C}_\alpha),\end{aligned}$$

as $\mathcal{F}_{n-1}$ preserves pullbacks (wide pullbacks).

Assume we know that M_r preserves pullbacks (wide pullbacks). Then, in dimension n

$$\lim_\alpha(M_{r+1}\mathcal{C}_\alpha)_n \simeq \lim_\alpha(A_n M_r \mathcal{C}_\alpha)_n \simeq (A_n M_r \mathcal{C})_n \simeq (M_{r+1}\mathcal{C})_n.$$

Now, in case (i),

$$\begin{aligned}W_n\mathcal{F}_n\mathcal{C} &\simeq \mathrm{colim}_r(M_r\mathcal{C}) \simeq \mathrm{colim}_r(\lim_\alpha(M_r\mathcal{C}_\alpha)) \\ &\simeq \lim_\alpha(\mathrm{colim}_r(M_r\mathcal{C}_\alpha)) \simeq \lim_\alpha W_n\mathcal{F}_n(\mathcal{C}_\alpha)\end{aligned}$$

as finite limits in $Glob_n$ commutes with filtered colimits.

In case (ii) we can not use the above calculation but we have the additional property that the unit of A_n is a cartesian natural transformation. Using the usual induction on n we can assume that $\mathrm{tr}_{n-1} W_n\mathcal{F}_n\mathcal{C} \simeq W_{n-1}\mathcal{F}_{n-1}\,\mathrm{tr}_{n-1}\mathcal{C}$ is isomorphic to $\mathrm{tr}_{n-1}\lim_\alpha(W_n\mathcal{F}_n\mathcal{C}_\alpha)$. It remains to establish this isomorphism in dimension n.

But in dimension n every morphism

$$(p_r)_n : (A_n M_r)_n \longrightarrow (M_{r+1})_n$$

is an isomorphism. And, for every projection $pr_\alpha : C_\alpha \to C_0$, we have a pullback

$$\begin{array}{ccc} (M_r\mathcal{C}_\alpha)_n & \xrightarrow{\epsilon_n} & (A_n M_r \mathcal{C}_\alpha)_n \\ \downarrow{pr_\alpha} & & \downarrow{pr_\alpha} \\ (M_r\mathcal{C}_0)_n & \xrightarrow{\epsilon_n} & (A_n M_r \mathcal{C}_0)_n \end{array}$$

Hence, for every α, r and k we have a pullback

$$\begin{array}{ccc} (M_r\mathcal{C}_\alpha)_n & \xrightarrow{c_{r,r+k}} & (M_{r+k} \mathcal{C}_\alpha)_n \\ \downarrow{pr_\alpha} & & \downarrow{pr_\alpha} \\ (M_r\mathcal{C}_0)_n & \xrightarrow{c_{r,r+k}} & (M_{r+k} \mathcal{C}_0)_n \end{array} \tag{4.1}$$

We now construct a map

$$\lim_\alpha \operatorname{colim}_r (M_r\mathcal{C}_\alpha)_n \to \operatorname{colim}_r \lim_\alpha (M_r\mathcal{C}_\alpha)_n$$

inverse to the canonical map. Indeed, let

$$\{m_\alpha\} \in \lim_\alpha \operatorname{colim}_r (M_r\mathcal{C}_\alpha)_n.$$

Here $m_\alpha \in \operatorname{colim}_r (M_r\mathcal{C}_\alpha)_n$. Moreover, as the unit of A_n is a monomorphism (because ϵ_n is cartesian) we know there exists a minimal finite $r(\alpha)$ such that $m_\alpha \in (M_{r(\alpha)}\mathcal{C}_\alpha)_n$. And, in addition, for $r_{\alpha_1} \leq r_{\alpha_2}$

$$c_{r(\alpha_1),r(\alpha_2)} \cdot pr_{\alpha_1} m_{\alpha_1} = pr_{\alpha_2} m_{\alpha_2}.$$

Using the pullback square (4.1) we see that r_{α_1} must be equal to r_{α_2}. This, therefore, determines an element from $\lim_\alpha (M_{r(\alpha)}\mathcal{C}_\alpha)_n$ and, finally, we have an element from $\operatorname{colim}_r \lim_\alpha (M_r\mathcal{C}_\alpha)_n$. The map thus constructed is obviously inverse to the canonical map. Hence, part (ii) is proved. □

THEOREM 4.2. *If A_n is a truncable cartesian finitary monad then the monad $\mathfrak{a}_n$ on $Comp_n$, induced by the adjunction $\mathcal{W}_n \dashv \mathcal{F}_n$, is cartesian.*

Let us prove, firstly, the following lemma:

LEMMA 4.2. *Let*

$$r'_n = r_n \cdot \mathcal{F}_n : \mathcal{F}_n \mathcal{W}_n \mathcal{F}_n \longrightarrow \mathcal{F}_n,$$

$$\Theta'_n = \Theta_n \cdot \mathcal{F}_n : V_n \mathcal{W}_n \mathcal{F}_n \longrightarrow W_n \mathcal{F}_n.$$

Under the conditions of the theorem, the natural transformations r'_n, Θ'_n are cartesian.

PROOF. It is easy to check for a truncable monad that A_n is cartesian implies A_k is cartesian. Hence, we can use induction.

For $k = 0$ the lemma is obviously true. Let us suppose it true for $k = n - 1$. We prove firstly, that Θ'_n is cartesian.

It is sufficient to check it for $(n-1)$-truncation and in dimension n separately. But $\mathrm{tr}_{n-1}\,\Theta'_n$ by definition is

$$W_{n-1}r_{n-1} \cdot \mathrm{tr}_{n-1}\,\mathcal{F}_n : W_{n-1}\mathcal{F}_{n-1}\mathcal{W}_{n-1}\,\mathrm{tr}_{n-1}\,\mathcal{F}_n \longrightarrow W_{n-1}\,\mathrm{tr}_{n-1}\,\mathcal{F}_n,$$

which is, by corollary 3.1, isomorphic to

$$W_{n-1}r'_{n-1}\,\mathrm{tr}_{n-1} : W_{n-1}\mathcal{F}_{n-1}\mathcal{W}_{n-1}\mathcal{F}_{n-1}\,\mathrm{tr}_{n-1} \longrightarrow W_{n-1}\mathcal{F}_{n-1}\,\mathrm{tr}_{n-1},$$

and, hence, is cartesian by the inductive assumption.

Now, consider a category P which has as objects the pairs (k, σ), where k is a natural number between 0 and $n-1$ and σ is one of the symbols $+$ or $-$. We have also the generating morphisms

$$s_- : (k, -) \longrightarrow (k-1, -),\ t_- : (k, -) \longrightarrow (k-1, +)$$

$$s_+ : (k, +) \longrightarrow (k-1, -),\ t_+ : (k, +) \longrightarrow (k-1, +)$$

and relations

$$t_- \cdot s_- = t_+ \cdot t_-\ ,\ s_- \cdot s_+ = s_+ \cdot t_+.$$

Then we can construct two functors $F, G : \mathit{Comp}_{n-1} \longrightarrow \mathit{Set}^P$:

$$F(\mathcal{C})(k, -) = F(\mathcal{C})(k, +) = (W_{n-1}\mathcal{F}_{n-1}\mathcal{W}_{n-1}\mathcal{F}_{n-1}\,\mathrm{tr}_{n-1}\,\mathcal{C})_k$$

$$G(\mathcal{C})(k, -) = F(\mathcal{C})(k, +) = (W_{n-1}\mathcal{F}_{n-1}\,\mathrm{tr}_{n-1}\,\mathcal{C})_k$$

and the values on s_-, s_+, t_-, t_+ are the corresponding source and target morphisms. We have also a natural transformation $f : F \to G$ which is equal to $W_k r'_k$ on the objects $(k, +)$ and $(k, -)$. By the previous argument f is cartesian.

We have, in addition, the functor $B(\mathcal{C}) = (W_n\mathcal{F}_n\mathcal{C})_n$ and a cone $\phi : B(\mathcal{C}) \longrightarrow G(\mathcal{C})$ given by the source and target functions.

We are, therefore, in the conditions of lemma 4.1. It means that we have the cartesian natural transformation

$$f^* : L \longrightarrow B.$$

It is not difficult to check that f^* coincides with the n-dimensional component of Θ'_n.

It remains to prove lemma 4.2 for r'_n. Recall that r'_n is the mate of $id : \mathcal{W}_n\mathcal{F}_n \to \mathcal{W}_n\mathcal{F}_n$. We can, therefore, apply the inductive construction of proposition 3.2 to the identity transformation and check that it gives a cartesian natural transformation

$$H_r : M_r\mathcal{W}_n\mathcal{F}_n \longrightarrow W_n\mathcal{F}_n$$

at every step. Then the colimiting natural transformation will be cartesian as well.

For $r = 0$ we see that H_0 coincides with Θ'_n and, hence, is cartesian. Suppose we have already proved that H_r is cartesian. By the construction of H_{r+1} we have to lift, firstly, H_r to a A_n-algebra morphism $\bar{H}_r$. But

$$\bar{H}_r = k \cdot A_n H_r$$

where k is the A_n-algebra structure morphism for $W_n\mathcal{F}_n$. It is cartesian because the multiplication in A_n is cartesian and k is determined by the colimit of the squares (3.3). Hence, $\bar{H}_r$ is cartesian.

Now, $\mathrm{tr}_{n-1} H_{r+1}$ is isomorphic to r'_{n-1} and $(H_{r+1})_n = (\bar{H}_r)_n$ (see the construction in 3.2)). Hence, H_{r+1} is cartesian and the lemma is proved. □

PROOF OF THE THEOREM 4.2. (a). The monad $\mathcal{a}_n$ preserves finite pullbacks by theorem 4.1(i).

(b). The multiplication in $\mathcal{a}_n$ is cartesian, because it is obtained as $\mathcal{W}_n r'_n$.

(c). There is a natural transformation

$$q : V_n \longrightarrow W_n \mathcal{F}_n$$

induced by the coprojection $M_0 \longrightarrow \mathrm{colim}_r M_r$. It is cartesian, because all the transformations $M_r \longrightarrow M_{r+1}$ are cartesian. The unit ν of the monad $\mathcal{a}_n$ mates with the identity transformation for $\mathcal{F}_n$ and, hence, q fits commutatively into the diagram

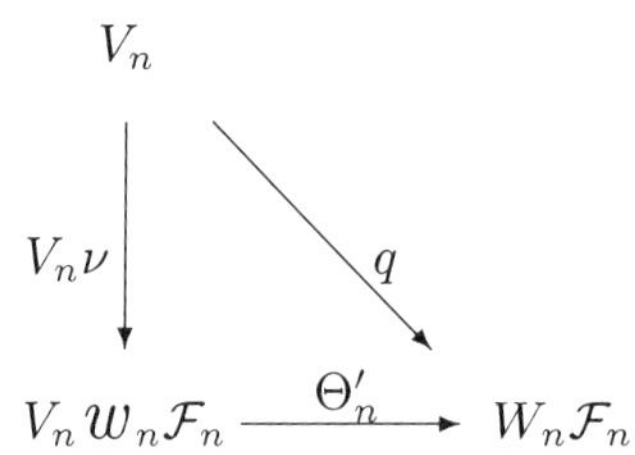

(see the construction of proposition 3.2). So $V_n \nu$ is cartesian. Using the inductive assumption we get that $\mathrm{tr}_{n-1} \nu$ is cartesian. Hence, ν is cartesian and we have finished the proof of the theorem. □

COROLLARY 4.1. *For an n-operad $\mathcal{A}_n$ in $Span_n$ the category of $\mathcal{A}_n$-computads is a preasheaf topos and the corresponding monad $\mathcal{a}_n$ is cartesian.*

PROOF. By a result of R. Street [**17**] every such monad is cartesian and preserves wide pullbacks. □

5. Presentation of algebras via computads.

THEOREM 5.1. *Let A_n be a truncable, cartesian monad on $Glob_n$ then the functor $\mathcal{W}_n : Alg_n \to Comp_n$ is monadic.*

We use for this the Beck's monadicity theorem in the form given in [**2**]. So we need to check that

1. $\mathcal{W}_n$ has a left adjoint;
2. $\mathcal{W}_n$ reflects isomorphisms;
3. Alg_n has coequalizers of $\mathcal{W}_n$-contractible pairs and $\mathcal{W}_n$ preserves them.

We have shown already that $\mathcal{W}_n$ has a left adjoint. The second condition can be easily verified using induction. So it remains to check the third.

LEMMA 5.1. *The natural transformation Θ_n has a right inverse*

$$\iota_n : W_n \to V_n \mathcal{W}_n.$$

PROOF. We will construct ι_n by induction. Define $\iota_0 = id$. Assume that ι_{n-1} is already defined and $\Theta_{n-1} \cdot \iota_{n-1} = id$. Let B be an A_n-algebra. Then

$$\mathrm{tr}_{n-1} W_n B = W_{n-1} \mathrm{tr}_{n-1} B$$

and

$$\operatorname{tr}_{n-1} V_n \mathcal{W}_n B = W_{n-1}\mathcal{F}_{n-1}\mathcal{W}_{n-1} \operatorname{tr}_{n-1} B$$

by definition. Then define $\operatorname{tr}_{n-1} \iota_n$ to be the following composite

$$W_{n-1} \operatorname{tr}_{n-1} B \xrightarrow{\iota_{n-1}} V_{n-1}\mathcal{W}_{n-1} \operatorname{tr}_{n-1} B \xrightarrow{V_{n-1}\nu_{n-1}} V_{n-1}\mathcal{W}_{n-1}\mathcal{F}_{n-1}\mathcal{W}_{n-1} \operatorname{tr}_{n-1} B$$

$$\xrightarrow{\Theta_{n-1}} W_{n-1}\mathcal{F}_{n-1}\mathcal{W}_{n-1} \operatorname{tr}_{n-1} B$$

where ν_{n-1} is the unit of the adjunction $\mathcal{F}_{n-1} \dashv \mathcal{W}_{n-1}$.
In dimension n we define

$$\iota_n(x) = (\iota_{n-1}s_{n-1}x, x, \iota_{n-1}t_{n-1}x).$$

Now we only need to prove that

$$W_{n-1}r_{n-1} \cdot \operatorname{tr}_{n-1} \iota_n = id,$$

where r_{n-1} is the counit of the adjunction $\mathcal{F}_{n-1} \dashv \mathcal{W}_{n-1}$ (see the definition of Θ_n).
Consider the following diagram

$$\begin{array}{ccccc}
V_{n-1}\mathcal{W}_{n-1} \operatorname{tr}_{n-1} B & & \xrightarrow{\quad V_{n-1}\nu_{n-1} \quad} & & V_{n-1}\mathcal{W}_{n-1}\mathcal{F}_{n-1}\mathcal{W}_{n-1} \operatorname{tr}_{n-1} B \\
 & \searrow^{id} & & \swarrow_{V_{n-1}\mathcal{W}_{n-1}r_{n-1}} & \\
\downarrow \Theta_{n-1} & & V_{n-1}U_{n-1} \operatorname{tr}_{n-1} B & & \downarrow \Theta_{n-1} \\
 & \swarrow_{\Theta_{n-1}} & & & \\
W_{n-1} \operatorname{tr}_{n-1} B & & \xleftarrow{\quad W_{n-1}r_{n-1} \quad} & & W_{n-1}\mathcal{F}_{n-1}\mathcal{W}_{n-1} \operatorname{tr}_{n-1} B
\end{array}$$

In this diagram the top triangle commutes by a triangle equation for an adjunction. The inner quadrangle commutes by naturality. Hence, the whole diagram commutes as well. So we have

$$W_{n-1}r_{n-1} \cdot \operatorname{tr}_{n-1} \iota_n = W_{n-1}r_{n-1} \cdot \Theta_{n-1} \cdot V_{n-1}\nu_{n-1} \cdot \iota_{n-1} =$$

$$= \Theta_{n-1} \cdot \iota_{n-1} = id.$$

□

LEMMA 5.2. *Let*

$$f, g : X \to Y$$

be a $\mathcal{W}_n$-contractible coequalizer pair of morphisms of A_n-algebras. Then the pair is W_n-contractible.

PROOF. Let

$$s : \mathcal{W}_n Y \to \mathcal{W}_n X$$

be a contraction of $(\mathcal{W}_n f, \mathcal{W}_n g)$. Then a trivial verification shows that the following composite

$$W_n Y \xrightarrow{\iota_n} V_n \mathcal{W}_n Y \xrightarrow{V_n s} V_n \mathcal{W}_n X \xrightarrow{\Theta_n} W_n A.$$

determines a contraction of the pair $(W_n f, W_n g)$. □

Now let $d : Y \to C$ be a coequalizer of (f, g) which exists due to our finitary assumption. Hence, by Beck's theorem ,

$$W_n d : W_n Y \to W_n C$$

is a coequalizer of $(W_n f, W_n g)$.

Let $\delta : \mathcal{W}_n Y \to \mathcal{Z}$ be a coequelizer of $(\mathcal{W}_n f, \mathcal{W}_n g)$. Then we have the morphism

$$m : \mathcal{Z} \to \mathcal{W}_n C$$

of computads which makes commutative the following diagram.

$$\begin{array}{ccc} \mathcal{W}_n Y & \xrightarrow{\delta} & \mathcal{Z} \\ & {\scriptstyle \mathcal{W}_n d} \searrow \quad \swarrow {\scriptstyle m} & \\ & \mathcal{W}_n C & \end{array} \tag{5.1}$$

We can reformulate now our theorem in the following form:

m **is an invertible morphism of computads.**

The following four lemmas provide a proof of this statement.

LEMMA 5.3. *Let $F : A \longrightarrow B$ be a functor and let $G : B \longrightarrow A$ be its right adjoint. Suppose the unit*

$$\epsilon : I \longrightarrow GF$$

is a monomorphism. Then F reflects mono and epimorphisms.

PROOF. Let $f : X \to Y$ be such that Ff is an epimorphism. Then, for an arbitrary object Z of A, we have a commutative diagram of sets

$$\begin{array}{ccc} A(Y,Z) & \xrightarrow{\epsilon^*} & A(Y,GFZ) \\ {\scriptstyle f^*}\downarrow & & \downarrow{\scriptstyle f^*} \\ A(X,Z) & \xrightarrow{\epsilon^*} & A(X,GFZ) \end{array}$$

In this diagram the morphisms ϵ^* are monomorphisms, and the right-side f^* is isomorphic by adjunction to

$$B(FY, FZ) \xrightarrow{(Ff)^*} B(FX, FZ)$$

which is a monomorphism. Hence, f^* is monomorphism and f is an epimorphism.

Suppose that Ff is a monomorphism. Then, by adjunction, the induced morphism of sets

$$A(Z, GFX) \longrightarrow A(Z, GFY)$$

is a monomorphism for an arbitrary Z and, hence, GFf is a monomorphism. By naturality of ϵ we have

$$\epsilon \cdot f = GFf \cdot \epsilon.$$

Thus f is a monomorphism. □

LEMMA 5.4. *The functor V_n reflects isomorphisms.*

PROOF. Let

$$f : \mathcal{C} \longrightarrow \mathcal{D}$$

be a morphism of A_n-computads such that $V_n f$ is an isomorphism. We thus have

$$\mathrm{tr}_{n-1} V_n f \simeq W_{n-1} \mathcal{F}_{n-1} \mathrm{tr}_{n-1} f$$

is an isomorphism. But W_{n-1} reflects isomorphisms. Let us prove that $\mathcal{F}_{n-1}$ reflects isomorphisms too.

By theorem 4.2 the unit of the adjunction $\mathcal{F}_{n-1} \dashv \mathcal{W}_{n-1}$ is a cartesian natural transformation and, hence, is a monomorphism as $Comp_{n-1}$ has a terminal object.

By lemma 5.3 this implies that $\mathcal{F}_{n-1}$ reflects mono and epimorphisms. But by theorem 4.1 the category $Comp_{n-1}$ is an elementary topos and, hence, mono+epi implies iso in $Comp_{n-1}$, which finishes the proof. □

Using the induction we can assume now that $\mathrm{tr}_{n-1} V_n m$ is an isomorphism. So it remains to prove that $V_n m$ is an isomorphism in dimension n.

LEMMA 5.5. *$(V_n m)_n$ is a monomorphism.*

PROOF. Let $a, b \in (V_n \mathcal{Z})_n$ be such that $V_n m(a) = V_n m(b)$. Then

$$s_{n-1} a = s_{n-1} b \ , \ t_{n-1} a = t_{n-1} b \tag{5.2}$$

by the inductive assumption.

Let now

$$a' = \Theta_n \cdot \tau(a), \ b' = \Theta_n \cdot \tau(b)$$

where $\tau : V_n \mathcal{Z} \to V_n \mathcal{W}_n Y$ is right inverse to $V_n \delta$ (recall that δ is the coequalizer of a contractible pair and hence splits). Then we have

$$V_n m \cdot V_n \delta \cdot \tau(a) = V_n m \cdot V_n \delta \cdot \tau(b).$$

Using the definition of m (see the triangle (5.1)), we get

$$V_n \mathcal{W}_n d \cdot \tau(a) = V_n \mathcal{W}_n d \cdot \tau(b).$$

Applying Θ_n to both sides of the last equation and using naturality of Θ_n we get

$$W_n d \cdot \Theta_n \cdot \tau(a) = \Theta_n \cdot V_n \mathcal{W}_n d \cdot \tau(a) = \Theta_n \cdot V_n \mathcal{W}_n d \cdot \tau(b) = W_n d \cdot \Theta_n \cdot \tau(b).$$

So we have

$$W_n d(a') = W_n d(b'). \tag{5.3}$$

Suppose that $a' \neq b'$. Consider an n-globular set $\{0,1\}_n$ which has exactly one element in every dimension less then n and two different elements 0 and 1 in dimension n. Then there exists a unique map

$$\lambda : V_n \mathcal{C} \to \{0,1\}_n,$$

such that $\lambda(x) = 0$ for every $x \in (V_n \mathcal{C})_n, x \neq b$ and $\lambda(b) = 1$. Construct also the unique map

$$l : W_n Y \to \{0,1\}_n,$$

with $l(x) = 0, x \neq b'$ and $l(b') = 1$. It is clear now that

$$\lambda \cdot V_n \delta = l \cdot \Theta_n.$$

From surjectivity of Θ_n (lemma 5.1) we deduce that

$$l \cdot W_n f = l \cdot W_n g.$$

This means that there should exist a map $\eta : W_nC \to \{0,1\}_n$ such that $l = \eta \cdot W_n d$. But this is impossible because of (5.3). Hence, $a' = b'$. This means, that $\tau(a) = \tau(b)$ because of (5.2). But τ is monomorphism so $a = b$ and we have proved the lemma. □

LEMMA 5.6. *$V_n m$ is a split epimorphism.*

PROOF. From the triangle 5.1 we see that it is sufficient to prove that $V_n \mathcal{W}_n d$ is a split epimorphism. The following commutative square

$$\begin{array}{ccc} F_nW_nY & \xrightarrow{F_nW_nd} & F_nW_nC \\ \downarrow k & & \downarrow k \\ Y & \xrightarrow{d} & C \end{array}$$

(where k is the counit of the adjunction $W_n \dashv F_n$) shows that we need to exhibit a splitting of

$$V_n \mathcal{W}_n F_n W_n d \tag{5.4}$$

and

$$V_n \mathcal{W}_n k. \tag{5.5}$$

By lemma 5.2 $W_n d$ is a coequalizer of a contractible pair and therefore splits. Hence, (5.4) splits as well.

To exhibit a splitting of (5.5) we construct a natural transformation

$$e_n : \mathcal{W}_n \longrightarrow \mathcal{W}_n F_n W_n$$

which is a right inverse for $\mathcal{W}_n k$. It is not hard to do using induction.

Indeed, put e_0 to be the unit ϵ_0 of the monad A_0. As $\mathcal{W}_0 = W_0$ we see that it indeed is a spliting of $\mathcal{W}_0 k$.

Assume that we have already defined e_k up to dimension $n-1$ with an additional property that the diagram

$$\begin{array}{ccc} W_k\mathcal{F}_k\mathcal{W}_kC & \xrightarrow{W_kr_k} & W_kC \\ \downarrow W_k\mathcal{F}_k e_k & & \downarrow \epsilon_k \\ W_k\mathcal{F}_k\mathcal{W}_kA_kC & \xrightarrow{W_kr_k} & W_kA_kC \end{array}$$

commutes. Then define

$$e_n = (e, E) : \mathcal{W}_n C \longrightarrow \mathcal{W}_n A_n C$$

as follows. The morphism E of A_{n-1}-computads is the composite

$$\mathcal{W}_{n-1}\,\mathrm{tr}_{n-1}\,C \xrightarrow{e_{n-1}} \mathcal{W}_{n-1}A_{n-1}\,\mathrm{tr}_{n-1}\,C \xrightarrow{\mathcal{W}_{n-1}\beta} \mathcal{W}_{n-1}\,\mathrm{tr}_{n-1}\,A_nC$$

and the morphism e in dimension n is defined by

$$e(\xi, a, \eta) = (e'(\xi), \epsilon_n a, e'(\eta))$$

where $e' = (W_{n-1}\mathcal{F}_{n-1}E)_{n-1}$. It is not hard to verify now using the inductive assumption that we thus get the required computad morphism. So $\mathcal{W}_n k$ splits and 5.5 is proved. □

6. Generalized pasting and some examples

EXAMPLE 6.1. Let Cat_ω be the category of (strict) ω-categories. Then the forgetful functor

$$Cat_\omega \to Glob$$

has a left adjoint and is monadic [**3**]. The corresponding monad on $Glob$ was denoted by D_s. It is not hard to see that our notion of $(D_s)_n$-computad coincides with the notion of n-computad of R. Street [**12, 14**].

Let $\mathcal{D}_n$ be a functor of free $(D_s)_n$-computad. For an n-category K the counit of the adjunction $\mathcal{D}_n \dashv \mathcal{W}_n$

$$r_n : \mathcal{D}_n \mathcal{W}_n K \to K$$

was called in [**14**] the *pasting operation* for K. This pasting operation allows the calculation of the composite for any pasting diagram in K.

Let us return to the general case. By analogy with the above example we give the following definitions

DEFINITION 6.1. The pasting operation for an A_n-algebra G is the counit

$$r_n : \mathcal{F}_n \mathcal{W}_n G \to G$$

DEFINITION 6.2. The n-dimensional pasting scheme for a finitary monad A_n is an n-cell in $\mathcal{F}_n 1_n$, where 1_n is the terminal A_n-computad.

DEFINITION 6.3. The interpretation (or labelling) of an n-dimensional pasting scheme $c \in W_n \mathcal{F}_n 1_n$ in an A_n-algebra G is an n-cell $x \in W_n \mathcal{F}_n \mathcal{W}_n G$ such that $W_n \mathcal{F}_n !(x) = c$, where ! is the unique morphism from $\mathcal{W}_n G$ to 1_n. The value of x is $W_n r_n(x) \in (W_n G)_n$

EXAMPLE 6.2. The most interesting case for us is that of monads generated by higher order operads from [**3, 16**]. Recall that an n-operad is an n-collection $\{A_T\}_{T \in Tr^n}$ equipped with a monoid structure with respect to a 'substitutional' tensor product of n-collections. Every n-operad A generates a monad on $Glob_n$

$$A(X)_k = \coprod_{T \in Tr_k} A_T \times X^T$$

where the coproduct is taken in an internal globular sense (see [**3, 16**]). The category of algebras for this monad is isomorphic to the category of algebras of the operad.

Example 1 is also from this class of monads as $(D_s)_n$ is generated by the terminal n-operad M. The terminality of M implies the *pasting theorem* in this case: the result of pasting depends only on the shape of the diagram and not on the method of calculation (see also [**9**]).

EXAMPLE 6.3. Let A be the following 2-collection. It has just one point for the unique 0-tree. For the 1-stage tree T with n-leaves $A_T(1)$ has the set $\mathcal{S}_n$, where $\mathcal{S}$ is the free nonsymmetric operad generated by the pointed collection of sets which has exactly one element in each of the dimensions $0, 1$ and 2 and has no element in all higher dimensions. Finally, for any 2-stage tree T,

$$A_T(2) = A_{\partial T}(1) \times A_{\partial T}(1).$$

It was shown in [**3**] that this 2-collection has the structure of a 2-operad. The category of algebras in $Glob_2$ of this operad is isomorphic to the category of bicategories and strict lax-functors (i.e. the structure cells for such a lax-functor are identities).

So an A-computad consists of a 2-globular set C such that the 1-skeleton of C is isomorphic to the free A_1-algebra generated by a 1-globular set (i.e. directed graph) D. It is not hard to see that the typical 1-cell in $A_1 D$ is a chain of cells of D and identities together with a chosen binary bracketing. For example, we could have

$$a \xrightarrow{\alpha} b \xrightarrow{id} b \xrightarrow{\alpha} c \xrightarrow{\gamma} d \xrightarrow{id} d \xrightarrow{id} d \xrightarrow{\xi} e$$

with the following bracketing

$$((\xi \cdot id) \cdot id) \cdot (\gamma \cdot (\beta \cdot (id \cdot \alpha))).$$

Hence, the pasting diagram for a bicategory consists of a usual 2-dimensional pasting diagram P, an indication of the 1-dimensional identities in P, and a binary bracketing of the 1-dimensional source and target for every 2-cell in P. In this case the pasting composite of the diagram is well defined. So we have rediscovered a result of D. Verity [**19**].

EXAMPLE 6.4. Another useful monad considered in [**3**] is the monad on $Glob_3$ generated by the 3-operad G which has as algebras the Gray-categories [**6**]. The 2-dimensional skeleton of G is the monad $\overline{Ssq}$ whose algebras are sesquicategories (i.e. 2-categories without interchange law [**14**]). The 1-skeleton of G coincides with $(D_s)_1$. Hence, a G-computad consists of a 3-globular set C whose 2-skeleton is isomorphic to a free sesquicategory generated by a 2-computad (in the sense of [**14**]). This notion of 3-dimensional computad was used by M. McIntyre and T. Trimble in their calculus of progressive 3D-diagrams [**11**]. They called it a Gray-computad.

It is clear that the corresponding pasting diagram is the usual three-dimensional pasting diagram together with information about the order of horizontal composition of 2-cells. The existence of such pasting diagrams appropriate for pasting in a Gray category was recently conjectured by S. Crans [**5**].

References

[1] J. Baez and J. Dolan, *Higher dimensional algebra III: n-categories and the algebra of opetopes,* Adv. Math. **135** (1998), 145–206.

[2] M. Barr and C. Wells, "Toposes, triples and theories," Springer Verlag, Berlin, 1985.

[3] M.A. Batanin , *Monoidal globular categories as a natural environment for the theory of weak n-categories,* Adv. Math. **136** (1998), 39–103.

[4] A. Carboni and P. Johnstone, *Connected limits, familial representability and Artin glueing,* Math. Structures Comput. Sci. **5** (1995) 441–459.

[5] S. Crans, *Pasting potpourri,* talk in the Australian Category Seminar, 25 June 1997; http://www-math.mpce.mq.edu.au/~coact/abstracts/acs970625a.html

[6] R. Gordon, A.J. Power and R. Street, "Coherence for tricategories," Mem. Amer. Math. Soc. **117** (1995) no. 558.

[7] C. Hermida, M. Makkai and J. Power, *On weak higher dimensional catagories,* preprint, 1997.

[8] A. Joyal and R. Street, *The geometry of tensor calculus, I,* Adv. Math. **88** (1991), 55–112.

[9] M. Johnson, *The combinatorics of n-categorical pasting,* J. Pure Appl. Algebra **62** (1989), 211–225.

[10] G. M. Kelly, *A unified treatment of transfinite constructions for free algebras, free monoids, colimits, associated sheaves and so on,* Bull. Austr. Math. Soc. **22** (1980) 1–84.

[11] M. McIntyre and T. Trimble, *Surface diagrams for Gray-categories,* preprint, 1997.

[12] R. Street, *Limits indexed by category-valued 2-functors,* J. Pure Appl. Algebra **8** (1976), 149–181.

[13] R. Street, *The algebra of oriented simplexes,* J. Pure Appl. Algebra **43** (1986), 235–242.

[14] R. Street, *Categorical Structures,* in "Handbook of algebra," vol. 1, ed. M. Hazewinkel, North-Holland, Amsterdam, 1996, pp. 529–577.

[15] R. Street, *Parity complexes,* Cahiers Topologie Géom. Différentielle Catégoriques **32** (1991), 315–343; *Corrigenda,* idem. **35** (1994), 359–361.

[16] R. Street, *The role of Michael Batanin's monoidal globular categories,* these proceedings.

[17] R. Street, *The petit topos of globular sets,* Macquarie Math. Report 98/232, 1998.

[18] Z. Tamsamani, *Sur des notions de n-catégorie et n-groupoide non-strictes via des ensembles multi-simpliciaux,* doctoral thesis, Université Paul Sabatier, Toulouse, 1995.

[19] D. Verity, *Enriched categories, internal categories and change of base,* doctoral thesis, Fitzwilliam College, Cambridge, 1992; Macquarie Math. Reports 93/123, 1993.

School of MPCE, Macquarie University, North Ryde, NSW 2109, Australia
E-mail address: mbatanin@mpce.mq.edu.au

Contemporary Mathematics
Volume **230**, 1998

Braided n-categories and Σ-structures

Lawrence Breen

ABSTRACT. We associate to any braided 2-groupoid with vanishing intermediate homotopy group a principal bundle (or torsor) endowed with a so-called Σ-structure, and show that this is the natural generalization to the 2-category context of the familiar quadratic invariant describing a braided groupoid. The corresponding structures for higher braided n-groupoids are also examined. This leads to the concept of Γ_3-torsor pairs, which are in the same relation to cubic forms as torsors endowed with a Σ-structure are to quadratic ones. The discussion also covers the corresponding properties of braided 2- and n-stacks in groupoids.

1. Introduction

This text is one of a series in which we explore the commutativity conditions which may be imposed on tensor laws in monoidal n-categories. We showed in [**6**] that the structure of a monoidal groupoid $\mathcal{C}$ could be analyzed in terms of a simpler geometric object, the associated commutator biextension, which measures the obstruction to the full commutativity of the tensor law of $\mathcal{C}$. This commutator biextension, which mimics the well-known commutator map associated to a central extension of abelian groups, is endowed with a very strong anti-symmetry property, which we termed its alternating structure [**4**]. Such alternating biextensions are classified up to isomorphism by the groups $\mathrm{Ext}^1(L\Lambda^2 B, A)$, where B and A are a pair of abelian groups, and $L\Lambda^2 B$ is the (non-additive) derived functor of the second exterior power functor $\Lambda^2_{\mathbf{Z}}$. In [**6**], we explored the manner in which monoidal n-groupoids may be described for $n \geq 2$ in terms of higher analogs of alternating biextensions, related to the higher derived exterior powers $L\Lambda^i_{\mathbf{Z}} B$. We always supposed there, as we will here, that the only nonvanishing homotopy groups of the n-groupoids considered were the extremal ones, the group $\pi_0(\mathcal{C}) = B$ of isomorphism classes of objects and the group $\pi_n(\mathcal{C}) = A$ of self-arrows of the identity $(n-1)$-arrow, since without this hypothesis the associated geometric objects are no longer standard ones. Such n-groupoids, together with their monoidal structure, are entirely determined up to equivalence by the associated k-invariant of their geometric realization, an element which lives in one of the cohomology groups $H^{r+n}(K(B,r), A)$. The integer r is determined by the level of commutativity of the given tensor multiplication law on $\mathcal{C}$.

1991 *Mathematics Subject Classification.* Primary 18D10; Secondary 14K25, 19D23.

In the present text, we consider monoidal n-groupoids $\mathcal{C}$ endowed with a braiding structure, in other words those classified up to equivalence by an element of the cohomology groups $H^{n+2}(K(B,2),A)$. We address the problem of determining the obstruction to the full commutativity of such a monoidal structure. In this situation, which is more restrictive than that examined in [**6**] where no braiding axiom was postulated in $\mathcal{C}$, the role previously assigned to the alternating commutator map associated to a central extension is taken up by the map $\tau : \Gamma_2^{\mathbf{Z}} B \longrightarrow A$ introduced by A. Joyal and R. Street in [**18**], whose source is the second divided power of the $\mathbf{Z}$-module B. The latter associates to an object X of $\mathcal{C}$ whose class in B is x the automorphism $\tau(x)$ of $X \otimes X$ determined by the braiding, viewed as an element of A. In the first situation examined here, that of a braided 2-groupoid $\mathcal{C}$, this map τ is replaced by the (trivial) A-principal bundle (or torsor) on B whose fiber above an element $x \in B$ is the A-torsor of 2-arrows from the identity 1-arrow of $X \otimes X$ to the braiding 1-arrow $\tau(x)$. The additional structure which this A-torsor posesses asserts that it satisfies the theorem of the cube, and that it is symmetric (*i.e.* invariant in a strong sense under pull-back by the inverse map of B). Such A-torsors were said in [**3**] to be endowed with a Σ-structure. They are much more classical objects than the alternating biextensions encountered above, since they correspond in an algebro-geometric context, when A is the multiplicative group G_m, to symmetric polarizations of the group scheme B.

In the higher categorical situations, one is led to geometric objects associated to the higher divided power functors $\Gamma_i^{\mathbf{Z}}$. Since braided 3-categories do not yield any suprisingly new multilinear structures, we mainly consider those structures associated to braided 4-categories. To these 4-categories correspond somewhat exotic geometric structures, which may be thought of as higher analogs of the torsors with Σ-structures defined by braided 2-categories. While Σ-structure A-torsors on B were classified by the group $\mathrm{Ext}^1(L\Gamma_2 B, A)$ associated to the derived functor $L\Gamma_2 B$ of Γ_2, these new structures, which we have termed Γ_3-A-torsor pairs on B, are classified by the group $\mathrm{Ext}^1(L\Gamma_3 B, A)$. These Γ_3-torsor pairs therefore stand in the same relation to cubic forms as the Σ-structure torsors do to quadratic ones. We unravel their definition here, but do not give a complete proof of the fact that the structures in question are indeed classified by this Ext group. Instead, we limit ourselves to the very strong plausibility argument given by proposition 5.2.

While we will speak freely in this text of n-categories for rather large integers n, ($n \leq 5$) will not make explicit the requisite definitions of such n-categorical structures. Even when this is not explicitly stated, our n-categories will always be assumed to be n-groupoids, in other words categories in which, for each $i < n$, i-arrows are invertible up to an $i+1$-arrow, and n-arrows are strictly invertible. The progress recently made by various authors in achieving a workable definition of an n-category, which is reflected in other texts [**2**], [**23**] of these Proceedings, justifies in our opinion this choice of the n-category terminology for what are in fact specific homotopy types determined by 2-stage Postnikov fibrations. A reader unwilling to endorse this point of view should consider this text as providing a very explicit description of a natural filtration on certain cohomology groups $H^{n+2}(K(B,2),A)$. Nor will we make explicit the corresponding definition of an n-stack, and simply refer to [**5**] for an illustration of what we have in mind in the case $n = 2$. An n-stacks in groupoids with trivial non extremal homotopy sheaves corresponds to a simplicial 2-stage Postnikov simplicial sheaf, described up to equivalence by the hyper-cohomology groups $\mathbf{H}^{r+n}(K(B,r),A)$. Here the extremal homotopy sheaves

$\pi_i(\mathcal{C})$ associated to $\mathcal{C}$ for are defined in the standard manner by sheafification of the naive homotopy presheaves associated to $\mathcal{C}$. The drawback with such a definition is that much of the geometry underlying the idea of an n-category, or an n-stack, viewed as a family of objects together with an assortment of arrows recedes into the background. We refer to [**22**],[**17**] for an up to date discussion of the theory of n-stacks. In the present context, the theory which we develop for n-groupoids extends to n-stacks in groupoids (which we will henceforth simply refer to as n-stacks). The biextensions, Σ-structures, and other geometric structures which we encounter in the n-groupoid case still exist, but no longer posess global sections which trivialize the underlying torsors. For the rest the situation is unchanged, with the proviso that additional non trivial terms in the filtration of the cohomology must be considered, which correspond to non vanishing higher Ext groups.

2. Braided categories and stacks

We begin by reviewing the classification of braided groupoids. Such a category may be defined directly, as in [**18**], as a groupoid endowed with a monoidal structure for which there exist inverses for objects and endowed with braiding morphisms[1]

$$\tau_{X,Y} : YX \longrightarrow XY \tag{2.1}$$

compatible with the monoidal structure. It may also be viewed as a 3-groupoid whose set of objets and of 1-arrows are both reduced to a single element. The latter description makes it clear how to classify such categories. Let $\mathcal{C}$ be a braided groupoid. The group $\pi_0(\mathcal{C})$ of isomorphism classes of objects of such a braided groupoid $\mathcal{C}$ and the group $\pi_1(\mathcal{C})$ of automorphisms of the identity object I of $\mathcal{C}$ are both abelian, and are respectively denoted B and A. The monoidal structure on $\mathcal{C}$ determines for every object $X \in \mathcal{C}$, by left multiplication by X, an identification

$$(X \otimes -) : B = \mathrm{Aut}_{\mathcal{C}}(I) \longrightarrow \mathrm{Aut}_{\mathcal{C}}(X) \tag{2.2}$$

The nerve $\mathcal{X}$ of $\mathcal{C}$ is a homotopy commutative H-space, which admits a double delooping to a space $\mathcal{Y} = B^2\mathcal{X}$ whose only non-trivial homotopy groups are

$$\pi_i(\mathcal{Y}) = \begin{cases} B & i = 2 \\ A & i = 3 \end{cases}$$

Such a space is entirely described, up to equivalence, by its unique k-invariant, an element $k(\mathcal{Y})$ in the cohomology group $H^4(K(B,2), A)$. The entire H-space structure of $\mathcal{X}$, in other words the braided category structure of $\mathcal{C}$, can be recovered from $\mathcal{Y}$ by the double delooping $\mathcal{X} \sim \Omega^2\mathcal{Y}$, so that this invariant yields a complete classification for equivalence classes of braided categories with associated groups B and A.

Let us now examine the integral homology of $K(B,2)$ in low degrees. This computation, which is functorial in B and goes back to [**13**] II, yields the following result

$$H_i(K(B,2)) = \begin{cases} B & i = 2 \\ 0 & i = 3 \\ \Gamma_2(B) & i = 4 \end{cases} \tag{2.3}$$

[1] We henceforth will denote by XY, rather than by the more customary $X \otimes Y$, the product of two objects X and Y in a monoidal category $\mathcal{C}$. On the other hand we will use the additive notation for the composition law in B.

A convenient method for studying the integral homology of $K(B,2)$ in all degrees, along the lines of the computations for $K(B,1)$ in [**7**], consists in observing that the homology of $K(B,2)$ (with the algebra structure determined by the Pontrjagin product) is endowed with a natural divided power structure on even degree elements. The natural map $B \simeq H_2(K(B,2))$ determined by the fundamental class therefore extends to a functorial algebra homomorphism

$$\Gamma_*(B) \longrightarrow H_*(K(B,2)) \tag{2.4}$$

When the abelian group B is torsion-free, the entire homology of $K(B,2)$ is easy to compute. The following result, which is implicit in [**8**] (exposé 11, theorems 4 and 6), is explicitly stated in [**14**] (proposition 3.4).

LEMMA 2.1. *Let B be a torsion-free abelian group. The canonical map (2.4) is an isomorphism.*

PROOF. For such a B, the homology $H_*(K(B,1))$ of the group B is isomorphic to the exterior algebra $\Lambda^*(B)$ of the $\mathbf{Z}$-module B. A bar-construction (or Eilenberg-Moore spectral sequence) argument implies directly that

$$H_*(K(B,2)) \simeq \mathrm{Tor}^{\Lambda^* B}(\mathbf{Z},\mathbf{Z})$$

It is an immediate consequence of the existence of the Koszul complex that the latter Tor term is isomorphic, as required, to $\Gamma_* B$. □

The methods of [**7**] now allow us to compute the homology of $K(B,2)$. Let $B_* \longrightarrow B$ be a torsion-free resolution of B. This determines a simplicial resolution $K(B_*,0)$ of the constant simplicial group $K(B,0)$, as well as a torsion-free simplicial resolution $K(B_*,2)$ of the simplicial abelian group $K(B,2)$. By the previous lemma, the E_1 term of the homology spectral sequence associated as in [**7**] §1 to $K(B_*,2)$ can be identified with the simplicial abelian group $\Gamma_*(B_*,0)$, obtained by applying the functor Γ_* componentwise to $K(B_*,0)$. Since the homology of $K(B_*,2)$ coincides with that of $K(B,2)$, the following proposition is obtained.

PROPOSITION 2.2. *There exists a functorial spectral sequence*

$$E^2_{p,q} = L_p\Gamma_q(B,0) \Longrightarrow H_{p+2q}(K(B,2))$$

which degenerates at E^2.

The degeneracy of the spectral sequence was essentially proved in [**7**] (with a different description of the initial term). The latter is here the pth left (non-additive) derived functor of Γ_q, as defined in [**12**], applied to B placed in degree 0, which will henceforth be simply denoted $L_p\Gamma_q B$ rather than the morse standard $L_p\Gamma_q(B,0)$. In low degrees, this proposition immediately yields the computations (2.3), as well as the isomorphism

$$H_5(K(B,2)) \simeq L_1\Gamma_2 B \tag{2.5}$$

We refer to [**3**] §9 for a discussion of the functor $L_1\Gamma_2 B$, and its identification with the functor $R(B)$ which by [**13**] theorem 22.1 describes the group $H_5(K(B,2))$.

We now consider the cohomology of $K(B,2)$. It follows from the universal coefficient theorem and the computation (2.3) that

$$H^4(K(B,2),A) \simeq \mathrm{Hom}(\Gamma_2(B),A) \tag{2.6}$$

so that braided categories $\mathcal{C}$ with associated groups B and A are classified up to equivalence by quadratic maps $\tau_{\mathcal{C}} : B \longrightarrow A$. As observed by A. Joyal and R. Street

([**18**] theorem 3.3), this identification, which follows from the explicit description of cycles for $H_4(K(B,n))$ (as in [**13**] theorem 26.1), associates to a category $\mathcal{C}$ with a braiding 2.1 the map

$$\tau = \tau_{\mathcal{C}} : B \longrightarrow A \tag{2.7}$$

which sends an element $x \in B$ to the automorphism τ_{X_x,X_x} of $X_x X_x$, for some representative object $X_x \in \mathcal{C}$ of x. The fact that $\tau_{\mathcal{C}}$ is independent of this choice of representative, and that it is quadratic, follows directly from (2.6). It is, however, of some interest to verify directly the quadraticity of this map in geometric terms. To do this conveniently, we begin by assuming that the multiplication law in our braided category $\mathcal{C}$ is strictly associative, a situation which can always be achieved by replacing $\mathcal{C}$ by an equivalent strictly associative monoidal category $\mathcal{D}$, to which the braiding may be transported. In this new category, the commutative hexagons which determine the braided structure reduce to triangles such as the triangles

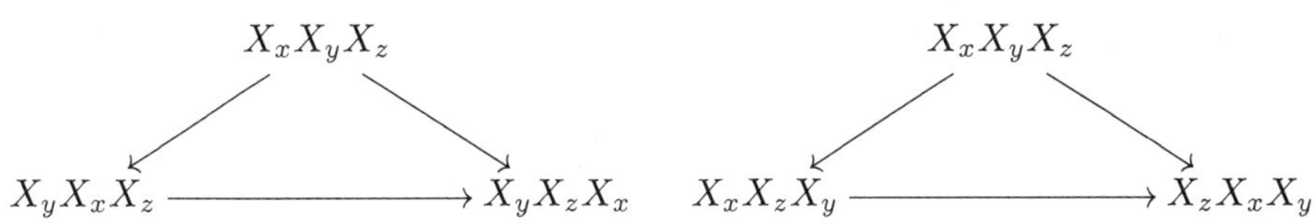

which respectively shuffle X_x from the left through $X_y X_z$ and X_z from the right through $X_x X_y$. We denote the first of these 'hexagon' diagrams by $H_{[x|_2\, y,z]}$ and the second by $H_{[x,y\,|_2\, z]}$ and observe that these notations are consistent with Eilenberg-Mac Lane's for cells in $K(B,2)$. Let us choose, for each pair of elements $x, y \in B$, an arrow

$$X_x X_y \xrightarrow{c_{x,y}} X_{x+Oy} \tag{2.8}$$

and henceforth denote, for any pair of elements $x, y \in B$, by $\tau_{x,y}$ the braiding map τ_{X_x,X_y} (2.1). We now consider the commutative diagram

(2.9)

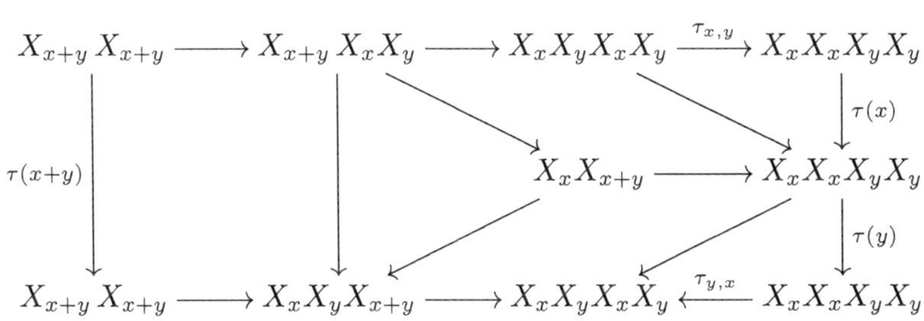

Here the middle triangle is $H_{[x+y|_2\, x,y]}$, and the top and bottom right-hand ones are respectively $H_{[x,y\,|_2\, x]}$ and $H_{[x,y\,|_2\, y]}$, the commutative squares being the obvious ones which follow from the functoriality of the maps τ. The commutativity of the outer square, when interpreted via arrows (2.8) as a square with the same object X_{2x+2y} at all vertices, asserts that the identity

$$\tau(x+y) - \tau(x) - \tau(y) = g(x,y) + g(y,x)$$

is satisfied in B, where $g(x, y) \in B$ is defined by the commutative diagram

$$\begin{array}{ccc} X_y X_x & \xrightarrow{\tau_{x,y}} & X_x X_y \\ {\scriptstyle c_{y,x}} \downarrow & & \downarrow {\scriptstyle c_{x,y}} \\ X_{x+y} & \xrightarrow[g(x,y)]{} & X_{x+y} \end{array} \tag{2.10}$$

Asserting that the map τ (2.7) is of degree two is therefore equivalent to showing that that the map $\phi(x, y) : B \times B \longrightarrow B$ defined by

$$\phi(x, y) = g(x, y) + g(y, x) \tag{2.11}$$

is bilinear. In order to prove this, consider first of all the well-known "Yang-Baxter" hexagon[2] associated to any three objects X, Y, Z in the braided category $\mathcal{D}$. This can be built up as follows from a pair of triangles $H_{[x|_2\, y,z]}$ and $H_{[x|_2\, z,y]}$

(2.12)

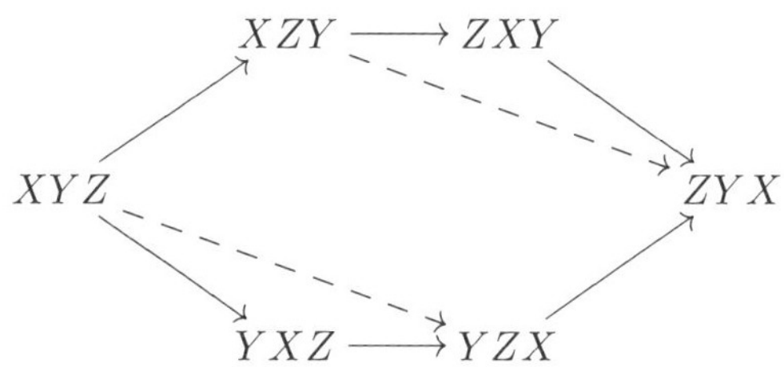

(the other method for constructing the identical Yang-Baxter hexagon is by the scheme

(2.13)

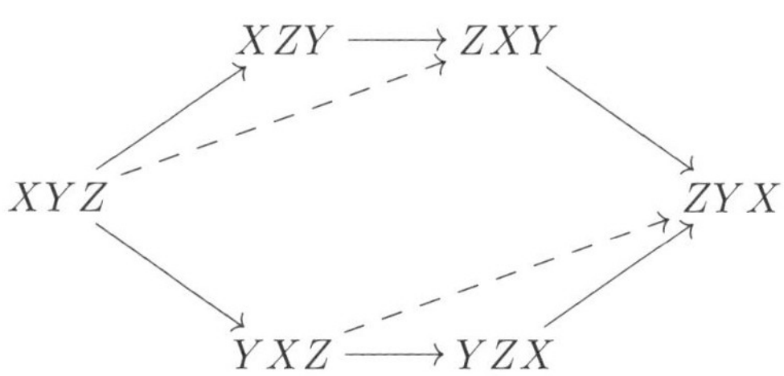

the upper and lower commutative triangles now being $H_{[x,y\,|_2\, z]}$ and $H_{[y,x\,|_2\, z]}$). In both instances, the commutativity of the central square follows from the functoriality of the braiding map (2.1)). Returning to the study of the map (2.11), its linearity in the first variable may be obtained by considering the following commutative diagram, built up out from the Yang-Baxter hexagon associated to the three objects X_x, X_y, X_z, to which have been glued a pair of triangles $H_{[x,z\,|_2\, y]}$, $H_{[y|_2\, z,x]}$ and a square whose commutativity follows, as always, from the functoriality of the

[2]This hexagon (which is actually a dodecagon when the associativity in the monoidal category is no longer assumed to be strict) is also referred to in the physics litterature as the "ABC = CBA" identity.

braiding map:

(2.14)

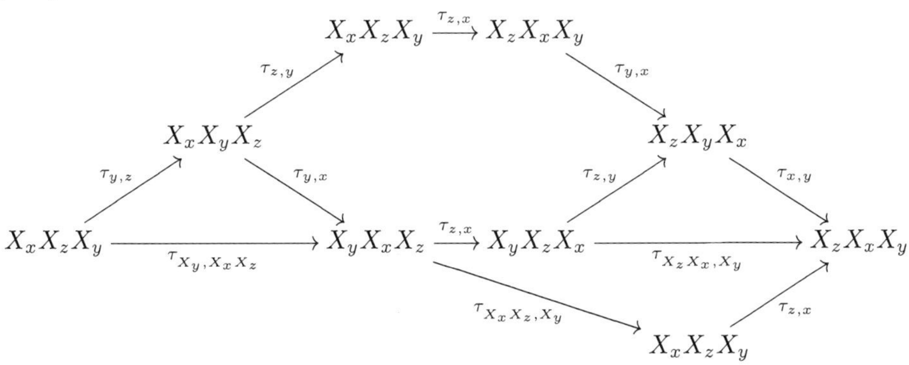

The arrows (2.8) allows us to transform each of the eight vertices of the outer diagram into the object X_{x+y+z}. The left hand (*resp.* right-hand) composite arrow then becomes our map $\phi(z,y)$ (*resp.* $\phi(x,y)$), and the lower composite one is $\phi(xz,y)$. The two final arrows $\tau_{z,x}$ encountered, when going around this diagram are both transformed into the maps $g_{x,z}$ (2.10), with opposite orientations. Since the group A of automorphisms of an object X in a braided category is abelian, these cancel out, and the required identity

$$\phi(xz,y) = \phi(x,y) + \phi(z,y)$$

is satisfied. The linearity of ϕ in the second variable y follows automatically, since the map ϕ is symmetric by construction.

In order to verify the quadraticity of the map τ (2.7), it is still necessary to ascertain that τ satisfies the identity

$$\tau(x) = \tau(-x) \tag{2.15}$$

Consider, for any $x \in B$, the commutative square

(2.16)

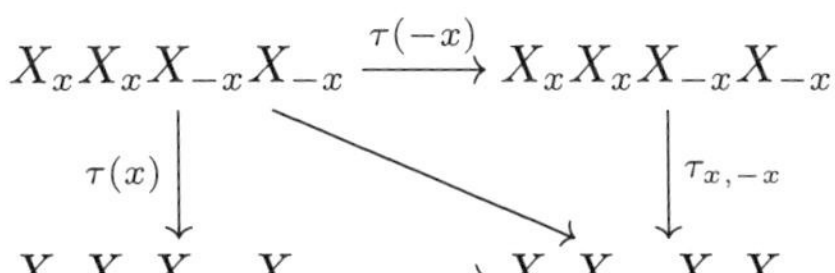

built from an upper triangle $H_{[x,-x\,|_2\,-x]}$ and a lower triangle $H_{[x|_2\,x,-x]}$. The common diagonal, when viewed after contraction to the identity object I of the middle term X_xX_{-x} at its source and one or the other of the expressions X_xX_{-x} at its target, is essentially the identity map from X_xX_x to itself. The triangles may therefore be viewed as having a common diagonal edge along which they may be attached as shown. By the process previously applied to the diagram (2.14), the outer square in (2.16) transforms to one in which all four vertices equal to $I = X_0$. The identity

$$\tau(x) + \phi(x,-x) = \tau(-x) + \phi(x,-x)$$

may be read off from its outer edges, so that the identity (2.15) is proved.

REMARK 2.3. *i*) In a category in which the multiplication law is not strictly associative, it is possible to carry out the above argument directly, without passing to a strictly associative equivalent category $\mathcal{D}$. A direct verification that $H_4(K(B,2)) \simeq \Gamma_2(B)$ makes it clear that the only significant pentagonal cell to be added in that case to diagram (2.9) is an associativity pentagon $P_{x,y,x,y}$, inserted at the common vertex $X_xX_yX_xX_y$ appearing in both the top and bottom position. A pentagon $P_{x,-x,x,-x}$ must similarly be added at the vertex $X_xX_{-x}X_xX_{-x}$ to diagram (2.16)

ii) It also follows from (2.3), and the universal coefficient theorem, that

$$H^3(K(B,2),A) \simeq \mathrm{Ext}^1(B,A) \tag{2.17}$$

$$H^2(K(B,2),A) \simeq \mathrm{Hom}(B,A) \tag{2.18}$$

and these relations have geometric interpretations analogous to that given above for (2.6). The first implies, as mentioned in [**18**] remark 3.3, that the equivalences $F : \mathcal{C} \longrightarrow \mathcal{D}$ between $\mathcal{C}$ and another braided category $\mathcal{D}$ equivalent to it, when viewed up to natural equivalence, form a principal homogenous space under the group $\mathrm{Ext}^1(B,A)$ of auto-equivalences of $\mathcal{C}$. The second relation asserts that the set of natural transformations between the identity equivalence $1_{\mathcal{C}} : \mathcal{C} \longrightarrow \mathcal{C}$ and some other fixed equivalence $u : \mathcal{C} \longrightarrow \mathcal{C}$ is either empty, or a principal homogeneous space under the group $\mathrm{Hom}(B,A)$ of natural transformations between $1_{\mathcal{C}}$ and itself. We will not make explicit the geometrical realization of these assertions, but observe that this yields, together with the interpretation given above of (2.6), a complete description up to 2-equivalence of the (symmetric monoidal) 2-category of braided categories with associated abelian groups B and A, analogous to the one given in [**18**] for the underlying 1-category.

The previous discussion may also be carried out in the more general context of a braided stack $\mathcal{C}$. B and A are now abelian groups in the topos T considered, with the proviso that a section x of B is only locally represented by an object X_x in $\mathcal{C}$. The isomorphisms (2.17) are essentially unchanged in this new situation, despite the fact that $H^i(K(B,2),A)$ now denotes the A-valued hypercohomology groups $\mathbf{H}^i(K(B,2),A)$ of the simplicial object $K(B,2)$ of T. The higher Ext^i groups no longer vanish for $i > 1$ in categories of abelian sheaves, as they did in the category of abelian groups, so that the isomorphism (2.6) must now be replaced by the exact sequence

$$0 \longrightarrow \mathrm{Ext}^2(B,A) \longrightarrow H^4(K(B,2),A) \longrightarrow \mathrm{Hom}(\Gamma_2(B),A) \longrightarrow \mathrm{Ext}^3(B,A) \tag{2.19}$$

which is a low degree consequence of the universal coefficient spectral sequence

$$\mathrm{Ext}^p(H_q(K(B,2)),A) \Longrightarrow H^{p+q}(K(B,2),A)$$

and of computation (2.3). The map $H^4(K(B,2),A) \longrightarrow \mathrm{Hom}(\Gamma_2(B),A)$ is essentially the one studied above. While it is *a priori* only defined locally, the fact that the explicit definition of the map (2.7) was independent of the choice of representing object X_x for a section $x \in B$ ensures that these local versions of $\tau_{\mathcal{C}}$ are compatible. They therefore glue in the stack context to a genuine, globally defined, sheaf map $\tau_{\mathcal{C}} : B \longrightarrow A$. While the map

$$H^4(K(B,2),A) \longrightarrow \mathrm{Hom}(\Gamma_2(B),A) \tag{2.20}$$

is no longer injective, its kernel $\mathrm{Ext}^2(B,A)$ once again has a natural interpretation as the group of equivalence classes of strictly symmetric monoidal categories with associated groups B and A [**11**]. The quadratic maps from B to A which live in the image of the map (2.20) thus still have, in the stack situation, a natural interpretation as the group of obstructions to the existence, on a braided monoidal category $\mathcal{C}$ with associated groups B and A, of a strictly symmetric structure. This may be directly understood by observing that, the vanishing of the map (2.7) implies that $\tau_{X,X}$ (2.1) is the identity map for all objects X in $\mathcal{C}$. It then follows from diagram (2.9) that for any pair of objects X and Y in $\mathcal{C}$ the composite map

$$YX \xrightarrow{\tau_{x,y}} XY \xrightarrow{\tau_{y,x}} YX$$

is also the identity map, so that $\mathcal{C}$ is indeed, as asserted, a strictly symmetric monoidal category.

3. Braided 2-categories and Σ-structures

Let us now pass from the study of categories to that of 2-categories. Braiding structures on such categories were introduced by Kapranov and Voevodsky [**19**] (see also [**1**], [**10**] definition 12, [**9**]). We restrict ourselves here to 2-groupoids (*resp.* 2-stacks) whose intermediate homotopy group $\pi_1(\mathcal{C})$ is trivial, in other words those for which, for any given pair of arrows $f, g : X \longrightarrow Y$ between a pair of objects X, Y, there exists (*resp.* there exists locally) a 2-arrow $u : f \Longrightarrow g$ connecting f to g. The first of the remaining homotopy groups of such a braided 2-groupoid $\mathcal{C}$ is the abelian group $\pi_0(\mathcal{C}) = B$ of isomorphism classes of objects of $\mathcal{C}$. For a given object X of $\mathcal{C}$ (which we may choose to be the identity object I of $\mathcal{C}$), the previous hypothesis ensures that the symmetric monoidal category $\mathcal{A} = \mathrm{Aut}_{\mathcal{C}}(X)$ is equivalent to the category with a unique object and an abelian group A of arrows. To make things specific, we may think of A as the group of 2-arrows

$$u : 1_I \Longrightarrow 1_I$$

from the identity 1-arrow $1_I : I \longrightarrow I$ of the identity object I to itself. In the stack case, both B and A are abelian sheaves, and $\mathcal{A}$ is equivalent as a symmetric monoidal stack to the stack $\mathrm{Tors}(A)$ of A-torsors. As discussed in [**5**] §8, the unique extra invariant necessary for a full description of $\mathcal{C}$ as a braided 2-category or stack is "Postnikov invariant" of $\mathcal{C}$, an element k in the cohomology (or hypercohomology group $H^4(K(B,2), \mathcal{A})$. Under our vanishing hypothesis for $\pi_1(\mathcal{C})$, this reduces to an element of the more traditional cohomology (or hypercohomology) group $H^5(K(B,2), A)$. A direct analysis of this group *via* the universal coefficient theorem is easy to achieve since the values of the requisite homology groups of B are know to us by (2.3) and (2.5). Let us consider instead, along the lines of the discussion carried out for monoidal 2-categories in [**6**], the filtration on the cohomology group $H^5(K(B,2), A)$ induced by the filtration on homology determined by proposition 2.2. In the so-called punctual case in which 2-categories (rather than to 2-stacks) are considered, the only term in the filtration which contributes something non trivial is the one involving Γ_2, so that one obtains an isomorphism

$$H^5(K(B,2), A) \simeq \mathrm{Ext}^1(L\Gamma_2 B, A) \tag{3.1}$$

Both left and right hand term can be subjected to a further anaysis via universal coefficient theorem. To the standard decomposition

$$0 \longrightarrow \mathrm{Ext}^1(H_4(K(B,2), A) \longrightarrow H^5(K(B,2), A) \longrightarrow \mathrm{Hom}(H_5(K(B,2)), A) \longrightarrow 0$$

of the left-hand term of (3.1) corresponds the decomposition

$$0 \longrightarrow \mathrm{Ext}^1(\Gamma_2(B), A) \longrightarrow \mathrm{Ext}^1(L\Gamma_2 B, A) \longrightarrow \mathrm{Hom}(L_1\Gamma_2(B), A) \longrightarrow 0$$

of the right-hand one, and this correspondence is consistent with the computation (2.3), (2.5) of the low degree homology of $K(B,2)$.

The term $\mathrm{Ext}^1(L\Gamma_2 B, A)$ which appeared in (3.1) may be interpreted geometrically as the group of isomorphism classes of A-torsors on B endowed with a so-called Σ-structure. For the reader's convenience we rapidly review this concept, and refer to [**3**] for additional information. It is a strengthening of the more familiar notion of a cube structure on such an A-torsor P on B. The latter is determined by a trivialization s of the "second difference" $\Theta(P)$ of the given torsor P, which is the A-torsor on B^3 defined by

$$\Theta(P) = m_{123}^* P \wedge \Big(\bigwedge_{ij} m_{ij}^* P\Big)^{-1} \wedge \Big(\bigwedge_{i} p_i^* P\Big) \tag{3.2}$$

where $\wedge$ is the contracted product of A-torsors, and the maps m_{123}, m_{ij}, p_i from B^3 to B are respectively the iterated sum, the partial (i,j)th sum and the projection on the ith factor. In addition, the trivialization s is required to satisfy a cocycle condition, for which we refer to [**3**] §2. An alternate method for describing a cube structure on the A-torsor P is to introduce its "first difference" of the A-torsor P. This is the A-torsor $\Lambda(P)$ on B^2 defined by

$$\Lambda(P) = m^*(P) \wedge p_1^*(P)^{-1} \wedge p_2^*(P)^{-1} \tag{3.3}$$

(where m is the group law and p_i are the projections from $B \times B$ to B). The A-torsor P is endowed with a cube structure if and only if $\Lambda(P)$ is not simply an A-torsor above B^2 but actually a biextension[3] of $B \times B$ by A.

A Σ-structure on an A-torsor P endowed with a cube structure is determined by the additional requirement that the torsor P be symmetric, in other words that there exists an isomorphism of A-torsors on B

$$\lambda : P \longrightarrow i^* P \tag{3.4}$$

(where $i : B \longrightarrow B$ is the inverse map for the group law of B) compatible in an appropriate sense with the cube structure of P (see [**3**] §5).

Example 3.1. In an algebro-geometric context, when A is the multiplicative group G_m and B a commutative group variety, an A-torsor on B is nothing else than a line bundle on B. In particular, when the group B is an abelian variety, we are assured of the existence of a trivialization s of the induced line bundle $\Theta(L)$ by the classical theorem of the cube [**20**]. In that situation, the cocycle condition mentioned above is satisfied, so that any such line bundle is automatically endowed with a cube structure. Furthermore, a symmetrization isomorphism λ (3.4) is then always compatible with the cube structure, whenever λ respects the rigidification, in other words sends the point $s(e,e,e)$ in the fibre P_e of P above the unit element $e \in B$ to itself.

Before explaining the correspondence (3.1) between braided 2-groupoids and Σ-structures in more geometric terms, let us examine in purely cohomological terms

[3]We refer to [**16**] for a definition of this concept.

the corresponding situation in the 2-stack case. The identification (3.1) is now replaced by an exact sequence

$$(3.5) \qquad \longrightarrow \mathrm{Hom}(\Gamma_2(B), A) \longrightarrow \mathrm{Ext}^3(B, A) \to H^5(K(B,2), A) \xrightarrow{\tau} \mathrm{Ext}^1(L\Gamma_2 B, A) \longrightarrow \mathrm{Ext}^4(B, A)$$

which prolongs the exact sequence (2.19). The term $\mathrm{Ext}^1(L\Gamma_2 B, A)$ once more describes A-torsors endowed with Σ-structures above B, but in the richer, sheaf-theoretic context in which a torsor does not, as in the punctual case, always posess a global section. The exactness of this sequence shows that the Σ-structure $\tau(\mathcal{C})$ associated by the middle map τ to a braided 2-stack $\mathcal{C}$ is trivial if and only if $\mathcal{C}$ is Picard strict, in the terminology of [**5**] definition 8.5.

This Picard strict definition is the strongest possible commutativity which can be imposed on a 2-stack, and it is of little interest in the punctual case since the group $\mathrm{Ext}^3(B, A)$ which classifies such structures for abstract abelian groups B and A vanishes. Let us briefly review from [**5**], for the reader's convenience, the relations between this and other possible commutativity conditions on 2-categories. This Picard strict commutativity condition is strictly stronger than the condition on monoidal 2-categories which we called "strongly involutory", which produces the 2-categories which R. Day and R. Street refer to as "symmetric Gray monoids" [**10**], and S. Crans as "symmetric monoidal 2-categories" [**9**]. The latter terminology for such 2-categories, which are classified under the vanishing hypothesis for $\pi_1(\mathcal{C})$ by elements in the group $H^7(K(B,4), A)$, seems the most appropriate one since the cohomology group in question lies in the stable range. This condition is in turn more restrictive than the condition which we refer to as "weakly involutory", and which applies to those categories which Day and Street call "sylleptic Gray monoids" and Crans refers to as "sylleptic monoidal 2-categories". The latter are classified, when $\pi_1(\mathcal{C})$ vanishes, by elements in $H^6(K(B,3), A)$. Finally, the commutativity conditions defining the original braided monoidal 2-categories of [**19**] are even weaker, since the latter are classified by elements of the group $H^5(K(B,2), A)$. In each of these instances the universal coefficient theorem yields, in the stack case, an edge-homomorphism map

$$\mathrm{Ext}^3(B, A) \longrightarrow H^{i+3}(K(B,i), A)$$

and these maps are compatible with the suspension maps

$$H^{i+3}(K(B,i), A) \xrightarrow{S} H^{i+2}(K(B, i-1), A)$$

as i varies.

If we now reason in geometric rather than homological terms, the Σ-structure associated to a braided 2-stack $\mathcal{C}$ may be understood by introducing the A-torsor $\mathcal{S}_{\mathcal{C}}$ on B, whose sections above an section $x \in B$ are the 2-arrows

$$(3.6) \qquad X_x X_x \underset{\tau(x)}{\overset{1_{X_x X_x}}{\rightrightarrows}} X_x X_x \qquad (\Downarrow S_x)$$

The A-torsor structure on $\mathcal{S}_{\mathcal{C}}$ is defined by composing such a locally defined 2-arrow with a 2-arrow from the identity map $1_{X_x X_x}$ to itself. In order to verify that $\mathcal{S}_{\mathcal{C}}$ is endowed with a Σ-structure, we must first verify that it is endowed with a cube structure, in other words that its "first difference"

$$\Lambda(\mathcal{S}_{\mathcal{C}}) = \mathrm{Isom}(p_1^*\mathcal{S}_{\mathcal{C}} \wedge p_2^*\mathcal{S}_{\mathcal{C}}, m^*\mathcal{S}_{\mathcal{C}})$$

is a biextension of $B \times B$ by A. By definition, a local section of $\Lambda(\mathcal{S}_{\mathcal{C}})$ is a rule which assigns to every (x, y) in some open set of $B \times B$, a triple of 2-arrows (S_x, S_y, S_{x+y}) (3.6), in a manner consistent with the indicated action of A. Reinterpreting diagram (2.9) in a 2-categorical context, we observe that the commutative triangles $H_{[x+y|_2\, x,y]}$, $H_{[x,y\,|_2 x]}$ and $H_{[x,y\,|_2 y]}$ have now been replaced by the corresponding hexagon 2-arrows determined by the braiding structure (see [**19**]). The other more trivial commutative cells in diagram (2.9) are also replaced by corresponding 2-arrows, so that diagram (2.9) determines, by composition of all these 2-arrows, a composite 2-arrow[4]

$$(3.7)\qquad \begin{array}{ccccc} X_{x+y}X_{x+y} & \longrightarrow & X_xX_yX_xX_y & \xrightarrow{\tau_{x,y}} & X_xX_xX_yX_y \\ & & & & \big\downarrow{\scriptstyle \tau(x)} \\ {\scriptstyle \tau(x+y)}\big\downarrow & & \swarrow\!\!\!\swarrow & & X_xX_xX_yX_y \\ & & & & \big\downarrow{\scriptstyle \tau(y)} \\ X_{x+y}X_{x+y} & \longrightarrow & X_xX_yX_xX_y & \xleftarrow{\tau_{y,x}} & X_xX_xX_yX_y \end{array}$$

Composing it with the given 2-arrows (S_x, S_y, S_{x+y}) according to the scheme indicated in the diagram

$$(3.8)\qquad \begin{array}{ccccc} X_{x+y}X_{x+y} & \longrightarrow & X_xX_yX_xX_y & \xrightarrow{\tau_{x,y}} & X_xX_xX_yX_y \\ & & & & {\scriptstyle \tau(x)}\big\downarrow \Leftarrow \big\downarrow{\scriptstyle 1} \\ {\scriptstyle 1}\big\downarrow \Rightarrow \big\downarrow{\scriptstyle \tau(x+y)} & & \swarrow\!\!\!\swarrow & & X_xX_xX_yX_y \\ & & & & {\scriptstyle \tau(y)}\big\downarrow \Leftarrow \big\downarrow{\scriptstyle 1} \\ X_{x+y}X_{x+y} & \longrightarrow & X_xX_yX_xX_y & \xleftarrow{\tau_{y,x}} & X_xX_xX_yX_y \end{array}$$

this yields a corresponding family of 2-arrows

$$(3.9)\qquad X_yX_x \;\underset{\tau_{y,x}\,\tau_{x,y}}{\overset{1_{X_yX_x}}{\rightrightarrows}}\; X_yX_x \qquad \Downarrow S_{x,y}$$

Let us now reconsider, in the 2-categorical context, the "Yang-Baxter" diagram (2.14), where all cells have now been replaced by 2-arrows determined by the braiding structure of $\mathcal{C}$. It may be viewed as a rule which assigns to every pair of

[4] This 2-arrow is roughly half of the diagram of 2-arrows denoted $((\bullet \otimes \bullet) \otimes (\bullet \otimes \bullet))$ in [**19**], when specialized from diagrams with vertices of the general form $XYZW$ and appropriate permutations thereof to those for which $X = Z$ and $Y = W$.

2-arrows (3.9)

(3.10)

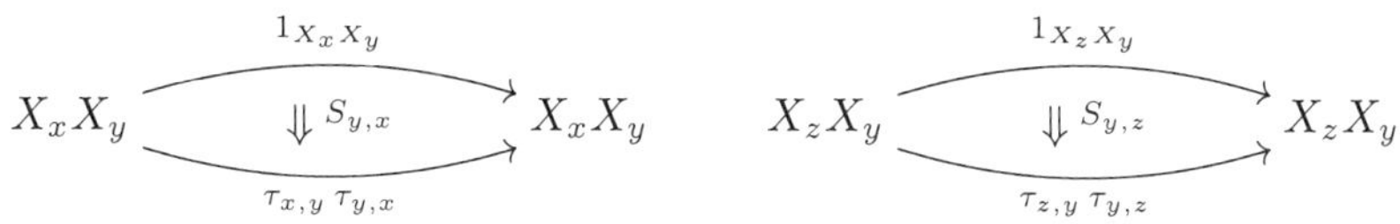

a corresponding 2-arrow

(3.11)

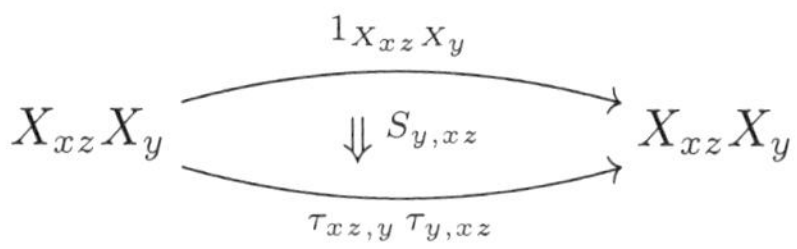

according the following scheme

(3.12)

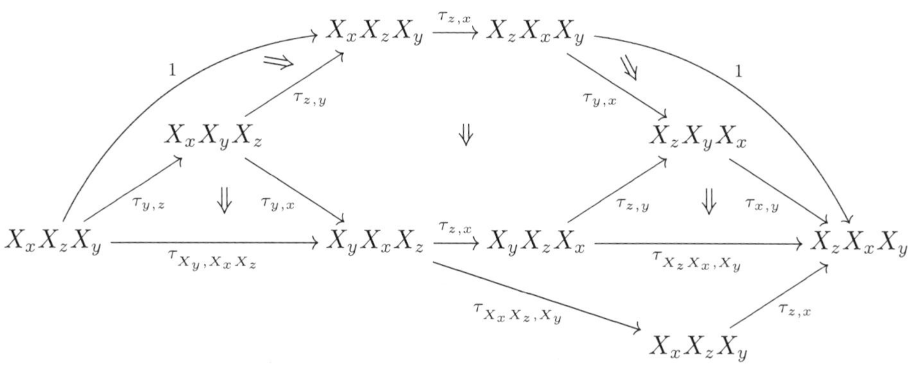

Such a rule may be viewed as the second partial composition law in the biextension $\Lambda(\mathcal{S}_{\mathcal{C}})$, so that what remains to be verified is the associativity of this law[5]. The commutativity is obtained by reversing some arrows and then reading diagram (3.11) from right to left. The associativity asserts that for any given triple of 2-arrows $S_{y,w}, S_{y,z}, S_{y,x}, S_{y,w}$ constructed as above from given the corresponding set of 2-arrows (3.6), the two possible methods for constructing, by the method of (3.12), the corresponding 2-arrow $S_{y,wxz}$

(3.13)

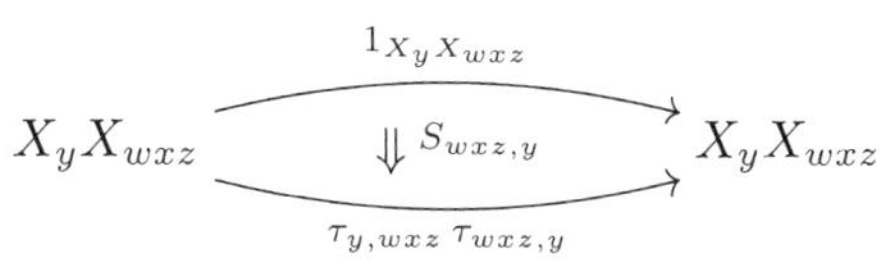

coincide. This is a compatibility condition for the corresponding diagrams (3.12) which we do not display here.

It remains to verify that the cube object $\mathcal{S}_{\mathcal{C}}$ is actually endowed with a Σ-structure. The requisite isomorphism (3.4) between $\mathcal{S}_{\mathcal{C}}$ and its pullback $i_B^* \mathcal{S}_{\mathcal{C}}$ by the inverse map i_B of B is easy to construct. It is a rule which associates to a section S_{-x} of the torsor $\mathcal{S}_{\mathcal{C}}$ above $-x$ a section which lives above the opposite

[5]According to [**3**] proposition 2.6, the partial group laws on a symmetric biextension $\Lambda(L)$ associated to an A-torsor L on B endowed with a cube structures are automatically compatible. Only the commutativity and the associativity conditions need to be verified.

point x of B. This is determined by the pasting scheme

(3.14)

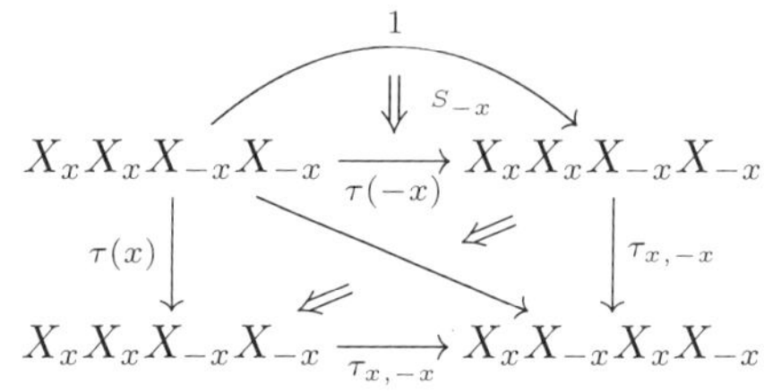

We do not attempt here a direct verification of the compatibility between this rule and the "Yang-Baxter" rule, even though this is part of the definition of a Σ-structure [**3**].

REMARK 3.2. It is now easy to understand in what sense $\mathcal{S}_{\mathcal{C}}$ is an obstruction to commutativity of the group law of $\mathcal{C}$: when $\mathcal{S}_{\mathcal{C}}$ is trivial, there exists a global family of 2-arrows S_x. The axioms for a strict Picard 2-category, or 2-stack, as given in [**5**] definition 8.5, may now be read off as follows from the previous Σ-structure conditions. The requirement that S_x be additive with respect to x, in the sense briefly described in [**5**] page 150, corresponds in our context to the assertion that the family of 2-arrows $S_{x,y}$ (3.9) coincide with that determined by the braiding structure on $\mathcal{C}$. The extra requirement that the 2-arrow displayed in [**5**] definition 8.5 coincide with $S_{x,x}$ is equivalent to the assertion above that the composite 2-arrow associated by (3.14) to a given 2-arrow S_{-x} is equal to S_x. The geometric content of the previous discussion may thus be summarized in very compact form as the assertion that a braided 2-stack endowed with a family of 2-arrows S_x satisfying the two condition just mentioned, is a strict Picard 2-stack, *i.e.* automatically satisfies the intermediate 'Picard' commutativity axioms for a symmetric monoidal 2-stack. In particular, this asserts in the category context that a braided 2-category endowed with a family of 2-arrows S_x satisfying these conditions (and for which $\pi_1(\mathcal{C}) = 0$) is equivalent to the trivial symmetric monoidal 2-category determined by a pair of abelian groups B and A.

4. Comparison with the non-commutative obstructions

It was shown in [**6**] §8 how to associate to any monoidal 2-category (or 2-stack) $\mathcal{C}$ determined as above by a pair of abelian groups B and A a so-called weak (2,2)-extension $\mathcal{E}$ of $B \times B$ by A, which under an additional condition was found to be alternating in a requisite sense. By the discussion at the very end of [**6**], a trivialization of $\mathcal{E}$ compatible with this entire structure makes $\mathcal{C}$ into a strict Picard stack (and therefore in the punctual case trivializes $\mathcal{C}$ compatibly with its monoidal structure). We now compare this (2,2)-extension obstruction $\mathcal{E}$ (which we will denote by $\mathcal{E}_{\mathcal{C}}$ when we wish to emphasize its dependence on the given 2-category $\mathcal{C}$) with the corresponding Σ-structure obtruction $\mathcal{S}_{\mathcal{C}}$ arising under the stronger assumption that $\mathcal{C}$ is braided.

Before passing to this 2-category situation, it is convenient to examine the corresponding assertion for monoidal and braided 1-categories. As recalled in §1, the obstruction to commutativity in a braided category or stack $\mathcal{C}$ with associated invariant groups B and A is determined by the quadratic map $\tau_{\mathcal{C}}$ (2.7) from B to A. On the other hand, by [**6**] proposition 3.1, there corresponds to any monoidal category or stack $\mathcal{C}$ with associated invariants B and A an A-torsor $E_{\mathcal{C}} = E$ above

$B \times B$, defined above $(x, y) \in B \times B$ by

$$E_{x,y} = \mathrm{Isom}_{\mathcal{C}}(X_y X_x, X_x X_y) \tag{4.1}$$

and which is endowed with the structure on a weak biextension[6]. Furthermore, when B is commutative, the associated commutator map, which examines whether the partial group laws on E are commutative, is a homomorphism $\phi_{\mathcal{C}}$ from $\Lambda^3 B$ to A, whose vanishing ensures that E is a (standard) biextension. The biextension E is then automatically an alternating one, a concept for which we refer to [**6**] §2, [**4**]. By [**6**] corollaries 5.3 and 6.1, E then measures the obstruction to the existence on $\mathcal{C}$ of a strict Picard structure. When the category $\mathcal{C}$ is braided, the relation between the quadratic map $\tau_{\mathcal{C}}$ and the alternating biextension $E_{\mathcal{C}}$ goes as follows

PROPOSITION 4.1. *Let $\mathcal{C}$ be a braided category with associated abelian groups B and A. The weak biextension $E_{\mathcal{C}}$ (4.1) associated to $\mathcal{C}$ is a genuine biextension, and in fact a trivial one. The quadratic map which by* [**4**] *proposition 1.5 determines the alternating structure on this trivial biextension E is, up to sign, the map $\tau_{\mathcal{C}}$ (2.7).*

PROOF. The braiding maps $\tau_{x,y}$ (2.1) determine a section s of the A-torsor E (4.1) above $B \times B$. The hexagon axioms assert that this section is bimultiplicative, and therefore trivializes E as a biextension. This shows *a posteriori* that the weak biextension $E_{\mathcal{C}}$ is in fact a genuine one. An alternating structure on $E_{\mathcal{C}}$ is determined by a trivialization of ΔE which differs from the trivialization Δs by a quadratic map τ. The equation

$$\tau(x) \circ \tau_{x,x} = 1_{X_x X_x}$$

in $\mathrm{Isom}_{\mathcal{C}}(X_x X_x, X_x X_x)$ proves that the the quadratic map τ associated to the canonical alternating structure on E determined as in [**6**] §5 by the identity arrow $1_{X_x X_x}$ on $X_x X_x$ is the opposite of the braiding automorphism $\tau_{x,x}$ of $1_{X_x X_x}$. □

REMARK 4.2. This proposition translates in cohomological terms to the assertion that there exists, for any pair of abelian groups (or sheaves of abelian groups) B and A, a diagram

$$\begin{array}{ccccccc} 0 \longrightarrow & \mathrm{Ext}^2(B,A) & \longrightarrow & H^4(K(B,2),A) & \longrightarrow & \mathrm{Hom}(\Gamma_2(B),A) \\ & \| & & \downarrow S & & \downarrow \delta \\ & \mathrm{Ext}^2(B,A) & \longrightarrow & G & \longrightarrow & \mathrm{Ext}^1(L\Lambda^2 B, A) \end{array} \tag{4.2}$$

with exact lines, and for which the left-hand square commutes, and the right-hand one anti-commutes. Here G is the kernel of the composite map

$$H^3(K(B,1),A) \longrightarrow \mathrm{Hom}(H_3(B),A) \longrightarrow \mathrm{Hom}(\Lambda^3 B, A) \tag{4.3}$$

determined by the universal coefficient theorem and the iterated Pontrjiagin product map $\Lambda^3 B \longrightarrow H_3(B)$. The vertical maps S and δ are respectively the suspension map and the boundary map induced on Ext^1 by the distinguished triangle

$$L\Gamma_2 B \longrightarrow B \otimes B \longrightarrow L\Lambda^2 B \xrightarrow{+1} L\Gamma_2 B[1] \tag{4.4}$$

[6]A weak biextension is simply a (standard) biextension in which the pair of partial composition laws are no longer required to be commutative, see [**16**] exposé VII 2.10.1.

Finally the lower horizontal exact sequence is described in detail, in geometric terms, in [**6**]. Its constituents are low degree terms in the spectral sequence

$$E_2^{p,q} = \mathrm{Ext}^p(L\Lambda^q B, A) \Longrightarrow H^{p+q}(K(B,1), A)$$

associated in cohomology to the homological spectral sequence (1.10) of [**7**].

Passing now from categories to 2-categories, the situation is governed, when the intermediate homotopy group $\pi_1(\mathcal{C})$ of the monoidal 2-category $\mathcal{C}$ is trivial, by the diagram

$$\begin{array}{ccccccc} \longrightarrow & \mathrm{Ext}^3(B,A) & \longrightarrow & H^5(K(B,2),A) & \longrightarrow & \mathrm{Ext}^1(L\Gamma_2(B),A) \\ & \| & & \downarrow{\scriptstyle S} & & \downarrow{\scriptstyle \delta} \\ \longrightarrow & \mathrm{Ext}^3(B,A) & \longrightarrow & K & \longrightarrow & \mathrm{Ext}^2(L\Lambda^2 B,A) \end{array} \tag{4.5}$$

Let us briefly review the structure of the group $H^4(K(B,1), A)$ which classifies such 2-categories $\mathcal{C}$ [**6**] §8. This group is endowed with a 4-term filtration, so that its subgroup K displayed above can no longer be expressed (as the corresponding group G in diagram (4.2)), as the kernel of some edge-homomorphism analogous to the map (4.3). Instead, K may be described as follows: to any element of $H^4(K(B,1), A)$, representing a given 2-category $\mathcal{C}$, is associated a so-called weak (2,2)-extension $\mathcal{E}_{\mathcal{C}} = \mathcal{E}$, whose fiber above a section (x, y) of $B \times B$, is defined by the formula

$$\mathcal{E}_{x,y} = \mathrm{Isom}_{\mathcal{C}}(X_y X_x, X_x X_y)$$

together with appropriate pair of partial tensor laws determined by the hexagon 2-arrows in $\mathcal{C}$. For the element in question to live in the subgroup K of $H^4(K(B,1), A)$, it is necessary that a certain quadrilinear map $\psi_{\mathcal{C}} : \Lambda^4 B \longrightarrow A$, and an induced alternating triextension $\Phi_{\mathcal{C}} \in \mathrm{Ext}^1(L\Lambda^3 B, A)$ both be trivial. In that case $\mathcal{E}$ is endowed with a genuine alternating (2,2)-extension structure, so that it determines as required by diagram (4.5) an element in $\mathrm{Ext}^2(L\Lambda^2 B, A)$. As explained in [**6**] §8, the trivialization of this (2,2)-alternating biextension ensures that the monoidal 2-stack $\mathcal{C}$ is strict Picard (and so is equivalent to the trivial monoidal category determined by the pair of groups B and A in the 2-category case). Since such strict Picard stacks are classified by elements of the group $\mathrm{Ext}^3(B, A)$, the present discussion gives a complete geometric interpretation of the lower line of diagram (4.5). The upper line in the diagram was introduced in (3.5) above, where it was described in similar geometrical terms.

The comparison between the A-torsor with Σ-structure $\mathcal{S}_{\mathcal{C}}$ (3.6) associated to a braided 2-stack $\mathcal{C}$, and the alternating (2,2)-biextension $\mathcal{E}_{\mathcal{C}}$ associated to the underlying monoidal 2-stack may now be carried out along the same lines as in the corresponding discussion in the 1-category case. The given braiding morphisms $\tau_{x,y}$ on $\mathcal{C}$ determine a global section of the A-gerbe $\mathcal{E}_{\mathcal{C}}$. The axioms satisfied by the hexagon 2-arrows in $\mathcal{C}$ imply that this global section is compatible with the partial group laws on $\mathcal{E}_{\mathcal{C}}$, and therefore trivializes $\mathcal{E}_{\mathcal{C}}$ as a (2,2)-extension. In particular, $\mathcal{E}_{\mathcal{C}}$ is represented by a class which lives in the subgroup K of $H^4(K(B,1), A)$, since this is where genuine (as opposed to weak) (2,2)-biextensions live. The exactness of the sequence

$$(4.6) \quad \cdots \longrightarrow \operatorname{Ext}^1(L\Gamma_2(B), A) \longrightarrow \operatorname{Ext}^2(L\Lambda^2 B, A) \longrightarrow \operatorname{Ext}^2(B \overset{L}{\otimes} B, A) \longrightarrow \operatorname{Ext}^2(L\Gamma_2(B), A) \longrightarrow \cdots$$

associated to the distinguished triangle (4.4) makes it clear that the alternating structure on $\mathcal{E}_{\mathcal{C}}$ is determined by a trivialization of its restriction $\Delta\mathcal{E}_{\mathcal{C}}$ above the diagonal in B. Such a restriction to the diagonal of a (2,2)-extension $\mathcal{E}$ is an A-gerbe on B canonically endowed with a structure which we might call a 2-Σ structure, and which is classified up to equivalence by the group $\operatorname{Ext}^2(L\Gamma_2 B, A)$. The complete definition of such a 2-Σ structure may be obtained along the same lines as that of alternating (2,2)-extensions in [**6**], by examining the dévissage of this Ext^2 group associated to the distinguished triangle

$$L\mathrm{Sym}^2 B \longrightarrow L\Gamma^2 B \longrightarrow B \overset{L}{\otimes} \mathbf{Z}/2 \overset{+1}{\longrightarrow} L\mathrm{Sym}^2 B\,[1]$$

We do not work this out in detail here, since such a discussion was already carried out in a related context in [**6**] §8. We simply recall from [**6**] that the sought-after trivialization of $\Delta\mathcal{E}_{\mathcal{C}}$, which determines the alternating structure on $\mathcal{E}_{\mathcal{C}}$, is given by the identity map

$$1_{X_x X_x} : X_x X_x \longrightarrow X_x X_x$$

viewed as an object in $(\Delta\mathcal{E}_{\mathcal{C}})_x = \operatorname{Isom}(X_x X_x, X_x X_x)$ for a varying section $x \in B$. The A-torsor with Σ-structure which describes this alternating structure on the torsor $\mathcal{E}_{\mathcal{C}}$ (trivialized by the given braiding arrows $\tau_{x,y}$ in $\mathcal{C}$) is, up to a sign depending on our conventions, the A-torsor $\mathcal{S}_{\mathcal{C}}$ of (locally defined) 2-arrows S_x (3.6) which compare the induced trivialization $\tau(x)$ of $\Delta\mathcal{E}_{\mathcal{C}}$ with the canonical trivialization $1_{X_x X_x}$. It follows from this discussion that the right-hand square in diagram (4.5) commutes up to sign, where the vertical map δ is defined either cohomologically as in (4.6), or geometrically as above.

5. Higher invariants: the higher Σ-structure torsors

We now briefly examine the invariants associated to higher homotopy types. We think of the homotopy types described by the groups $H^6(K(B,2), A)$ as being represented by appropriately defined braided 3-categories $\mathcal{C}$ (or rather, 3-groupoids since we assume that every n-arrow in $\mathcal{C}$ is invertible up to an $n+1$-arrow) with the usual additional assumptions: the group $\pi_0(\mathcal{C})$ of isomorphism classes of objects is isomorphic to B, the group $\pi_3(\mathcal{C})$ of self 3-arrows of any identity 2-arrow is isomorphic to A, and the intermediate homotopy groups $\pi_i(\mathcal{C})$ are trivial. We will not spell out the braiding axioms here, but merely observe that one method for obtaining them would be to think of $\mathcal{C}$ as a 5-category with a single object and a single 1-arrow. The multilinear objects attached, as in the lower dimensional case, to an appropriate filtration of $H^6(K(B,2), A)$ are now the groups $\operatorname{Hom}(\Gamma_3 B, A), \operatorname{Ext}^2(L\Gamma_2 B, A)$ and $\operatorname{Ext}^4(B, A)$. We have seen that the group $\operatorname{Ext}^2(L\Gamma_2 B, A)$ describes classes of what might be termed 2-Σ structures, the higher (1-categorical) analog of the Σ-structure A-torsors on B. We do not spell out here the definition of such structures, nor of the strict Picard 3-categorical structures classified by $\operatorname{Ext}^4(B, A)$ since this is routine. Let us merely observe that 2-Σ-structures are to (ordinary) Σ-structures as their anti-symmetric analog, the alternating (2,2)-extensions of [**6**] are to (ordinary) alternating biextensions. The only essentially new feature which occurs when braided 3-categories are considered is thus the appearance of the degree three component $\Gamma_3 B$ of the divided power algebra $\Gamma_* B$ of B, together with the projection

$$H^6(K(B,2),A) \longrightarrow \mathrm{Hom}(\Gamma_3 B, A) \tag{5.1}$$

The latter map is readily understood from a topological point of view, since it is induced by the isomorphism $\Gamma_3 B \simeq H_6(K(B,2))$ [**13**][7] theorem 21.1 determined by the divided power algebra structure of $H_*(K(B,2)$, and by the universal coefficient theorem. In categorical terms, it can be understood by observing that in a braided 3-category, the two possible methods of construction of a Yang-Baxter 2-arrow described by diagrams (2.12, 2.13) no longer yield the same result, as in a braided 2-category. Instead, they can only be compared by a 3-arrow, which is part of the the braiding axioms for 3-categories, and which we denote, for representative objects X_x, X_y, X_z in $\mathcal{C}$ of the isomorphism classes $x, y, z \in B$, by $(YB)_{x,y,z}$. Such a 3-arrow corresponds to the degree 6 cell in Eilenberg-Mac Lane's chain complex for $K(B,2)$ traditionally denoted by $[\,x|_2\, y|_2\, z]$. It corresponds, in Getzler-Jones' cell decomposition [**15**] §5.4 of the quotiented configuration space $\overset{\circ}{F_3}(2)$ of pairs of points in $\mathbf{R}^3$, to the cell $[1\,||\,2\,||\,3]$. It now follows from the explicit computation of the elements of $H_6(K(B,2))$ that the map (5.1) associates to the given braided 3-category $\mathcal{C}$ the homorphism

$$\Upsilon_{\mathcal{C}} : \Gamma_3 B \longrightarrow A \tag{5.2}$$

which sends an element $\gamma_3(x)$ to the 3-arrow $(YB)_{x,x,x}$. The latter specialized Yang-Baxter 3-arrow is, as required an element of the group A of self 3-arrows in $\mathcal{C}$ from an appropriate trivial 2-arrow to itself.

The group $\Gamma_3 B$ differs from the previously encountered group $\Gamma_2 B$, since it is no longer generated by indecomposable elements of the form $\gamma_3(x)$, additional generators of the form $\gamma_2(x)y$ with $x, y \in B$ being required. The map (5.2) can therefore no longer be entirely described by the values of the elements $(YB)_{x,x,x} \in A$ for varying $x \in B$. Instead, one is also led to consider the composite map

$$\Gamma_2 B \otimes B \longrightarrow \Gamma_3 B \xrightarrow{\Upsilon_{\mathcal{C}}} A$$

determined by the algebra structure in $\Gamma_* B$. The latter may be viewed as a map

$$B \times B \longrightarrow A$$

which is quadratic in the first variable x and linear in the second variable y. It is constructed by assembling the three Yang-Baxter 3-arrows $(YB)_{x,x,y}$, $(YB)_{x,y,x}$, $(YB)_{y,x,x}$ into a self 3-arrow, in a manner which we will not make explicit here.

We now pass to the groups $H^7(K(B,2),A)$, whose elements describes the braided 4-categories with trivial intermediate homotopy groups. There are now three corresponding multilinear invariants, which live in the groups

$$\mathrm{Ext}^1(L\Gamma_3 B, A) \qquad \mathrm{Ext}^3(L\Gamma_2 B, A) \qquad \mathrm{Ext}^5(B, A)$$

While the last two represent some significant structure in the 4-stack case, they both vanish when 4-categories are considered, so we will restrict ourselves to a discussion of the geometric objects representing the remaining group $\mathrm{Ext}^1(L\Gamma_3 B, A)$. This is

[7] In [**13**], the standard degree i component $\Gamma_i B$ of the divided power algebra $\Gamma_* B$ is somewhat confusingly denoted $\Gamma_{2i} B$.

in any case the most interesting of the three, since it involves the degree three functor $\Gamma_3 B$. This functor lives in the exact sequence

$$\Gamma_2 B \otimes B \longrightarrow \Gamma_3 B \longrightarrow B/3B \longrightarrow 0 \tag{5.3}$$

reminiscent of the short exact sequence

$$0 \longrightarrow \mathrm{Sym}^2 B \longrightarrow \Gamma_2 B \longrightarrow B/2B \longrightarrow 0$$

by which one analyzes a Σ-structures in terms of the associated symmetric biextension [**3**] §8. In the present situation, the first arrow is the multiplication map considered above, and the second the map which sends $\gamma_3(x)$ to the class of $x \bmod (3B)$.

Since there exist two sorts of generators for $\Gamma_3 B$, the geometric structure classifed by the group $\mathrm{Ext}^1(L\Gamma_3 B, A)$ involves a pair of objects (E, L) where E an A-torsor on $B \times B$ and L is an A-torsor on B. When such a pair (E, L) arises from a braided 4-category $\mathcal{D}$ with trivial intermediate homotopy groups, the corresponding A-torsor $L_{\mathcal{D}}$ is defined by

$$L_{\mathcal{D}} = \mathrm{Isom}_{\mathcal{D}}(1, \Delta_{123}^*(YB))$$

Its fiber above $x \in B$ is the torsor of 4-arrows from the identity 3-arrow to $(YB)_{x,x,x}$. The fibre above $x, y \in B^2$ of the corresponding A-torsor $E_{\mathcal{D}}$ similarly consists of the torsor of 4-arrows in $\mathcal{D}$ from the identity 3-arrow to the 3-arrow assembled from $(YB)_{x,x,y}$, $(YB)_{x,y,x}$ and $(YB)_{y,x,x}$ to which we previously alluded (without making it explicit) in the 3-category case.

Returning to the general situation, we now describe the additional structure to be imposed on the pairs (E, L). The map from $\mathrm{Ext}^1(\Gamma_3 B, A)$ to $\mathrm{Ext}^1(L\Gamma_2 B \overset{L}{\otimes} B, A)$ induced the multiplication in $\Gamma_* B$ is the first projection from (E, L) to E, so that the first element of structure to be imposed on the pair (E, L) is that E be quadratic in the first variable and linear in the second. This means that for each $y \in B$, $E_{(-,y)}$, viewed as a torsor above B, is endowed with a Σ-structure, and that $E_{(x,-)}$, viewed for each $x \in B$ as an A-torsor on B, is a commutative extension of B by A. We also require the compatibility of these two structures, in other words that the partial multiplication map

$$E_{x,y_1} \wedge E_{x,y_2} \longrightarrow E_{x,y_1+y_2} \tag{5.4}$$

determined by the second variable group law respect the Σ-structures.

The next set of conditions on the pair (E, L) arises from the relations

$$\gamma_3(x + x') - \gamma_3(x) - \gamma_3(x') = \gamma_2(x)\, x' + \gamma_2(x')\, x \tag{5.5}$$

$$3\gamma_3(x) = \gamma_2(x)\, x \tag{5.6}$$

satisfied by the divided powers. The first of these relations translates into the requirement that there exists an isomorphism of A-torsors

$$\Lambda(L) \xrightarrow{\alpha} E \wedge s^* E \tag{5.7}$$

where s is the map which permutes the factors in B^2, and $\Lambda(L)$ is the "first difference" A-torsor of L, defined above B^2 formula (3.3). Similarly, the relation (5.6) translates to the requirement that there exists an isomorphism of A-torsors

$$\Delta E \xrightarrow{\beta} L^3 \tag{5.8}$$

between the restriction of E to the diagonal and the contracted third power $L \wedge L \wedge L$ of the torsor L. It is obvious that such a requirement is satisfied by the pair

$(E_{\mathcal{D}}, L_{\mathcal{D}})$ since sections of the restriction of $E_{\mathcal{D}}$ to the diagonal involve three copies of $(YB)_{x,x,x}$.

Consider the pullback

$$(\Lambda \times 1)^*(E \wedge s^*E) \xrightarrow{(\Lambda\times 1)^*\alpha} (\Lambda \times 1)^*(\Lambda(L))$$

by $(\Lambda \times 1)$ of the map α (5.7). The target of this map is the "first difference of the first difference " of L, which is canonically isomorphic to the "second difference" $\Theta(L)$ of L (3.2). By linearity of E in the first variable, the source of this map simplifies so that it reduces to an isomorphism

$$(\Lambda \times 1)^*E \longrightarrow \Theta(L) \tag{5.9}$$

Since E posesses a Σ-structure with respect to its first variable, and is linear with respect to the second one, the source of this map is canonically endowed with a triextension structure above $B \times B \times B$, and in fact one which is symmetric with respect to the interchange of the first two variables. In particular the A-torsor L is endowed *via* this isomorphism with a hypercube structure, since its second difference is multilinear[8].

We now list the conditions which the maps α and β must satisfy. Observe that the target $\Theta(L)$ of the map (5.9) is invariant up to canonical isomorphism under any permutation of the three factors in B^3. The canonical isomorphism $\Theta(L) \longrightarrow \sigma_{123}^*\,\Theta(L)$ induced by the cyclic permutation σ_{123} of the factors of B^3 transports to an isomorphism

$$\psi : (\Lambda \times 1)^*E \simeq \sigma_{123}^*((\Lambda \times 1)^*E)$$

It is convenient to consider this in the fiber above a general element $(x, y, z) \in B^3$. It can then be displayed as an isomorphism[9]

$$\psi_{x,y,z} : E_{x+y,z}\, E_{x,z}^{-1}\, E_{y,z}^{-1} \simeq E_{y+z,x}\, E_{y,x}^{-1}\, E_{z,x}^{-1} \tag{5.10}$$

Taking into account the maps $E_{z,x}\, E_{z,y} \simeq E_{z,x+y}$ and $E_{x,y}\, E_{x,z} \simeq E_{x,y+z}$ which expresses the linearity of E in the second variable, one therefore obtains an isomorphism ϕ between the requisite pullbacks of E, whose fiber $\phi_{x,y,z}$ above (x, y, z) is an isomorphism of A-torsors

$$E_{x+y,z}\, E_{z,x+y}\, E_{y,x}\, E_{x,y} \longrightarrow E_{y+z,x}\, E_{x,y+z}\, E_{z,y}\, E_{y,z} \tag{5.11}$$

We impose on the map α (5.7) the requirement that the following diagram commutes.

$$\begin{array}{ccc} (E_{x+y,z}\, E_{z,x+y})\,(E_{y,x}\, E_{x,y}) & \xrightarrow{\phi_{x,y,z}} & (E_{y+z,x}\, E_{x,y+z})\,(E_{z,y}\, E_{y,z}) \\ {\scriptstyle \alpha_{x+y,z}\,\alpha_{x,y}}\downarrow & & \downarrow{\scriptstyle \alpha_{x,y+z}\,\alpha_{y,z}} \\ \Lambda(L)_{x+y,z}\,\Lambda(L)_{x,y} & \longrightarrow & \Lambda(L)_{x,y+z}\,\Lambda(L)_{x,y} \end{array} \tag{5.12}$$

the lower horizontal map being the (obvious) canonical isomorphism. A second constraint on the map (5.7) is much simpler to state. It is the requirement that

[8] so that its third difference is trivial.

[9] The contracted product symbols $\wedge$ will henceforth mostly be omitted.

the following commutative square, in which the horizontal maps are the canonical ones, also commutes:

$$\begin{array}{ccc} \Lambda(L) & \xrightarrow{\simeq} & s^*\Lambda(L) \\ {\scriptstyle\alpha}\downarrow & & \downarrow{\scriptstyle s^*\alpha} \\ E \wedge s^*E & \longrightarrow & s^*E \wedge E \end{array} \tag{5.13}$$

A final axiom involves both the arrows α (5.7) and β (5.8). Observe first of all that the map ψ (5.10) specializes to

$$\psi_{y,y,x} : E_{2y,x}\, E_{y,x}^{-2} \simeq E_{x+y,y}\, E_{y,y}^{-1}\, E_{x,y}^{-1}$$

On the other hand, the given Σ-structure on the first variable of E determines an isomorphism [**3**] (6.3.2)

$$E_{2y,x} \simeq E_{y,x}^{4}$$

so that the map in question determines an isomorphism

$$E_{x+y,y} \simeq E_{x,y}\, E_{y,y}\, E_{y,x}^{2}$$

Exchanging the variables x and y this yields

$$E_{x+y,x} \simeq E_{y,x}\, E_{x,x}\, E_{x,y}^{2}$$

These two isomorphisms, and the linearity of E in the second variable therefore determine a composite map

$$E_{x+y,x+y} \simeq E_{x+y,x}\, E_{x+y,y} \simeq E_{x,x}\, E_{y,y}\, E_{x,y}^{3}\, E_{y,x}^{3}$$

which may also be viewed as a map

$$\Lambda\Delta E \xrightarrow{\gamma} E^3 \wedge s^*E^3 \tag{5.14}$$

The final condition which we impose on the pair (E, L) is the commutativity of the following diagram in which the top horizontal arrow is the canonical one

$$\begin{array}{ccc} \Lambda(L^3) & \xrightarrow{\simeq} & \Lambda(L)^3 \\ {\scriptstyle\Lambda(\beta)}\downarrow & & \downarrow{\scriptstyle\alpha^3} \\ \Lambda\Delta E & \xrightarrow{\gamma} & E^3 \wedge s^*E^3 \end{array} \tag{5.15}$$

We summarize the previous discussion as follows.

DEFINITION 5.1. Let A and B be two abelian groups (or sheaves of abelian groups). A Γ_3-A-torsor pair above B consists of an A-torsor L on B and an A-torsor E on $B \times \mathrm{B}$ satisfying the following conditions:

i) E is endowed with a Σ-structure with respect to the first variable and with a commutative extension structure with respect to the second. These conditions are compatible, in the manner previously described (5.4).

ii) There exists a pair of isomorphisms of A-torsors (5.7), (5.8)

$$\Lambda(L) \xrightarrow{\alpha} E \wedge s^*E \qquad\qquad \Delta E \xrightarrow{\beta} L^3$$

iii) The diagrams (5.12), (5.13), and (5.15) commute.

In order to obtain a complete proof that the category of such quadruples (E, L, α, β) is described by the truncated complex $t_{\leq 0}\mathrm{RHom}(L\Gamma_3(B), A[1])$, whose homology groups are the groups $\mathrm{Ext}^i(L\Gamma_3(B), A)$ for $i = 0, 1$, one would have to construct a simple realization of the object $L\Gamma_3(B)$, as was done in [**3**] §8 for the object $L\Gamma_2(B)$. We do not carry this out here, but observe the following fact, which provides, in view of the exactness of the sequence (5.3), very strong evidence for this assertion. For this reason, and for lack of a better name, we will call these quadruples "Γ_3-A-torsor pairs on B", or simply "Γ_3-torsor pairs"

PROPOSITION 5.2. *Let (E, L, α, β) be a Γ_3 torsor pair for which there exists a global section of E, compatible with both the Σ-structure of E in the first variable and with the linearity in the second one. Then L is endowed with a group law, for which it is a commutative extension of B by A. Furthermore the pullback of the extension*

$$0 \longrightarrow A \longrightarrow L \longrightarrow B \longrightarrow 0$$

by the map $B \xrightarrow{3} B$ is canonically split.

PROOF. The given trivialization of E induces by (5.7) a trivialization of the torsor $\Lambda(L)$, and this in turn determines a composition law on L which is compatible with the projection from L to B. The commutativity of diagram (5.12) implies that the induced trivialization of the source and target of the lower horizontal map in this diagram are compatible, a statement which is equivalent to the associativity of the corresponding composition law on L. Similarly, the commutativity of diagram (5.13) implies that the trivializations of the source and target of the upper horizontal map in the latter diagram are compatible, and this in turn is equivalent to the commutativity of the composition law on L. This shows that L is a commutative extension of B by A (since the existence of an inverse map on L, the existence of a unit element and the compatibility of the composition law in L with the inclusion of A in L follow automatically). Finally, the commutativity of diagram (5.15) implies that the trivialization of the A-torsor L^3 determined by the map β (5.8) is compatible with the group law on L^3 induced by the group law of L, and therefore splits the pushout L^3 of L by the map $A \xrightarrow{3} A$ as a group extension. This is equivalent to the assertion regarding the pullback of L by the map $B \xrightarrow{3} B$. □

References

[1] J. Baez and M. Neuchl, *Higher-dimensional algebra I: Braided Monoidal 2-Categories*, Adv. in Math. **121** (1996) 196-244.

[2] J. Baez and J. Dolan, *Categorification*, these Proceedings.

[3] L. Breen, *Fonctions thêta et théorème du cube*, Lecture notes in math. **980**, Springer Verlag (1983).

[4] L. Breen, *Biextension alternées*, Compositio math. **63** (1987) 99-122.

[5] L. Breen, *Classification of 2-gerbes and 2-stacks*, Astérisque **225**, Société mathématique de France (1994).

[6] L. Breen, *Monoidal categories and multiextensions*, to appear in Compositio Math.

[7] L. Breen, *On the functorial homology of abelian groups*, to appear in the J. of Pure and Applied Algebra.

[8] H. Cartan, Séminaire H. Cartan E.N.S. 7e année: 1954/55, *Algèbres d'Eilenberg-MacLane et Homotopie*, Secrétariat Mathématique, Paris (1956).

[9] S. Crans, *Generalized centers of braided and sylleptic monoidal 2-categories*, Advances in Math. **136** (1998), 183-223.

[10] B. Day and R. Street, *Monoidal bicategories and Hopf algebroids*, Advances in Math. **129** (1997), 99-157.

[11] P. Deligne, *La formule de dualité globale* exposé XVIII of *Théorie des Topos et Cohomologie Etale des Schémas* vol. 3 (SGA 4) Lecture notes in math. **305**, Springer Verlag (1973).

[12] A. Dold and D. Puppe, *Homologie nicht-additiven funktoren. Anwendungen* Ann. Inst. Fourier **11** (1961), 201-312.

[13] S. Eilenberg and S. Mac Lane, *On the groups $H(\Pi, n)$ I, II*, Ann. of Math. **58** (1953) 53-106, **60** (1954) 49-139.

[14] G. Decker *The integral homology algebra of an Eilenberg-Mac Lane Space*, Ph. D. thesis, University of Chicago (1974).

[15] E. Getzler and J.D.S Jones, *Operads, homotopy algebra, and iterated integrals for double loop spaces*, preprint.

[16] A. Grothendieck, *Séminaire de Géométrie Algébrique du Bois-Marie, 1967-69 (SGA7) I* Lecture notes in math. **288**, Springer-Verlag (1972).

[17] A. Hischowitz and C. Simpson, *Descente pour les n-champs*, preprint (1998), available as math.AG/9807049.

[18] A. Joyal and R. Street, *Braided tensor categories*, Advances in Math. **102** (1993) 20-78.

[19] M. Kapranov and V. Voevodsky, *2-categories and Zamolodchikov equations*, Proc. Symp. Pure Math. **56** (2) (1994) 177-259.

[20] D. Mumford, *Geometric Invariant Theory*, Ergebnisse, N.S. **34**, Springer-Verlag 1965.

[21] M. Neuchl, Ph.D. thesis, Munich (1997).

[22] C. Simpson, *A closed model structure for n-categories, internal Hom, n-stacks and generalized Seifert-Van Kampen*, preprint (1997), available as alg-geom/9704006.

[23] R. Street, *The role of Michael Batanin's monoidal globular categories*, these Proceedings.

UMR CNRS 7539, INSTITUT GALILÉE, UNIVERSITÉ PARIS 13, F-93430 VILLETANEUSE
E-mail address: breen@math.univ-paris13.fr

Contemporary Mathematics
Volume **230**, 1998

Categories of vector bundles and Yang-Mills equations

Jean-Luc Brylinski

ABSTRACT. We describe applications of gerbes and their differential geometry to various problems in gauge theory and geometric quantization. For instance, the study of quantum rank r vector bundles over a symplectic manifold leads to studying some flat bundles of matrix algebras. The point of view of gerbes leads to a modification of the Yang-Mills equation involving a source term. There is a close connection between gerbes and a vector bundle of the notion of 2-vector space, which we call 2-vector bundle. The problem of finding global objects of a given rank is analyzed in terms of degree 3 cohomology.

1. Introduction

Gauge theory leads to natural classes of connections on manifolds, for instance Yang-Mills connections. On the other hand, the differential geometry of gerbes developed in [**Bry**] leads to extensions of concepts of line bundles and connections over them. We study here some extensions of the theory of gerbes to higher rank bundles, and how they lead to interesting classes of bundles and connections.

The first example (§2) is that a rank r quantum vector bundle on a symplectic manifold (M, ω). By this we mean a vector bundle with connection whose curvature is $\omega \otimes Id$. We study the question of existence of such bundles in terms of a non-abelian gerbe (Theorem 2.2), and we obtain a concrete criterion in terms of flat Azumaya bundles.

In the other examples considered in §3 and §4, we consider some auxiliary fibration $E \to B$ and study some vector bundles over E with extra structures depending on the geometry of the fibration. Often, as in gauge theory, there will be a class of connections over the bundle, and some modification of the curvature forms will descend to Lie algebra valued 2-forms over B. We will find that over a riemannian four-manifold B the classical Yang-Mills equations satisfied by these 2-forms have to be modified by addition of a source term. There are two types of relevant fibrations: in §3 we study a family $E \to B$ of symplectic manifolds, and in §4 we consider a fibration whose fiber is an infinite-dimensional projective space.

It most often happens that vector bundles of the type we are interested exist locally over the base B, but not globally. The obstruction problem for finding a

1991 *Mathematics Subject Classification.* Primary 18G50, 58E15, 58F05; Secondary 16K20.
This research was supported in part by NSF grant DMS-9504522

global object leads to classes in non-abelian cohomology; in the rank one case, the class lives in $H^3(B,\mathbb{Z})$. This can be formulated either using gerbes or using 2-vector spaces. This is studied in §4, where we introduce a category over any open set of B whose objects are vector bundles of the type we are considering. By varying the open set, one gets a sheaf of categories over B; this leads to a relative version of the notion of 2-vector space of Kapranov-Voevodsky [**K-V**], which we call a 2-vector bundle. We establish in Proposition 5.6 the equivalence between 2-vector spaces and the abelian gerbes studied in [**Bry**]. We also study the relation between abelian and non-abelian gerbes.

2. Quantum vector bundles over symplectic manifolds

Before our first example, recall the notion of quantum line bundle over a symplectic manifold (M,ω). This means a line bundle L over M equipped with a connection ∇ whose curvature is equal to $2\pi i\omega$. We also have the notion of a quantum vector bundle: this means a complex vector bundle V over M with a connection ∇ whose curvature is equal to the scalar-valued 2-form $2\pi i\omega\otimes Id$. There can exist a quantum vector bundle of rank r even if there is no quantum line bundle. There is an easy necessary condition on ω: since $r\cdot\omega$ represents the first Chern class of V, we must have the integrality condition

$$r\cdot\omega\in Im[H^2(M,\mathbb{Z})\to H^2(M,\mathbb{C})]. \tag{2.1}$$

In dimension 2 from the work of Atiyah-Bott [**A-B**] we have

PROPOSITION 2.1. *Given a closed 2-form ω on the compact oriented surface M of genus ≥ 1, there exists a rank r quantum vector bundle over M if and only if $r\cdot\int_M\omega\in\mathbb{Z}$.*

PROOF. We just noted that the condition is necessary. Conversely, assume $r\cdot\int_M\omega=k\in\mathbb{Z}$. First assume that ω is the volume form corresponding to some riemannian metric. Then consider the (unique) smooth rank r vector bundle E over M with first Chern class k. According to [**A-B, Prop. 6.16**], there exists a central Yang-Mills connection ∇ on E. The curvature of this connection equal to $2\lambda\pi i\omega\otimes Id$ for some constant λ. Since the first Chern class of E is k, we have $\lambda r\int_M\omega=k$, hence $\lambda=1$. Therefore (E,∇) is a quantum vector bundle. Now for general ω there is a volume form ω' which is cohomologous to ω. Then there is a vector bundle with connection (E,∇) of rank r whose curvature is equal to $2\pi i\omega'\otimes Id$. On there other hand there is a line bundle with connection (L,D) with curvature $2\pi i(\omega-\omega')$. Then the tensor product $(E\otimes L,\nabla+D)$ has curvature $2\pi i\omega\otimes Id$. □

For a manifold M of dimension $2k$ the integrality condition for the higher Chern classes imposes extra constraints on an integer r such that there exists a quantum vector bundle of rank r, namely

$$\binom{r}{l}[\omega^l]\in H^{2l}(M,\mathbb{Z}),\ \forall l\in\{1,\cdots,r\} \tag{2.2}$$

We wish to prove a version of Proposition 2.1 in higher dimensions. We assume that ω satisfies (2.1). Let $\exp(2\pi i[\omega])$ denote the image of the class of $[\omega]\in H^2(M,\mathbb{C})$ by the exponential map $\exp:\mathbb{C}\to\mathbb{C}^*$. Then $\exp(2\pi i[\omega])$ is killed by r.

There is a systematic way of creating classes in $H^2(M, \mathbb{C}^*)$ killed by r. We recall that for any group G, flat G-bundles are classified by the non-abelian cohomology set $H^1(M, G)$, which is isomorphic to $H^1(\pi_1(M), G)$ for M connected. We are interested in flat $PGL(r, \mathbb{C})$-bundles, for which the central extension $1 \to \mathbb{C}^* \to GL(r, \mathbb{C}) \to PGL(r, \mathbb{C})$ yields an exact sequence due to Frenkel [**Fr**]:

$$\cdots \to H^1(M, GL(r, \mathbb{C}) \to H^1(M, PGL(r, \mathbb{C})) \xrightarrow{\delta} H^2(M, \mathbb{C}^*) \tag{2.3}$$

The map δ can be described as follows: let g_{ij} be a $PGL(r, \mathbb{C})$-valued Čech cocycle with respect to a geodesically convex open covering of M. Pick a lift $\tilde{g}_{ij} \in GL(r, \mathbb{C})$; then we have $\tilde{g}_{ik} = c_{ijk}\tilde{g}_{ij}\tilde{g}_{jk}$, for a 2-cocycle c_{ijk} with values in $\mathbb{C}^*$, which represents $\delta(g_{ij})$. Such a class is automatically r-torsion.

THEOREM 2.2. *Let ω be closed 2-form on M such that $r\omega$ is an integral cohomology class. Let* $\exp(2\pi i[\omega]) \in H^2(M, \mathbb{C}^*)$ *denote the image by the exponential map class of the class of $2\pi i\omega$ in $H^2(M, \mathbb{C})$. Then the following two conditions are equivalent:*

(1) There exists a rank r quantum vector bundle.

(2) The class $\exp(2\pi i[\omega])$ *is of the form $\delta(\eta)$ for some class*

$$\eta \in H^1(M, PGL(r, \mathbb{C})).$$

The proof will use the theory of gerbes in the sense of Giraud. So we begin by recalling some definitions. First we need the notion of a stack, or sheaf of categories over B. This means that for each open set U we have a category $\mathcal{C}_U$ and for $V \subset U$ we have a restriction functor $\mathcal{C}_U \to \mathcal{C}_V$, which composes well in triples $W \subset V \subset U$ of open sets. To have a sheaf of categories, we need two conditions. First of all, given two objects P, Q of $\mathcal{C}_U$, the presheaf $\underline{Hom}(P, Q)$ over U should actually be a sheaf; this is the gluing condition for morphisms. Then we need a gluing axiom for objects of $\mathcal{C}$, which roughly speaking, says that given an open covering (U_i) of B, objects P_i of $\mathcal{C}_{U_i}$ and isomorphisms $\phi_{ij} : P_j \tilde{\to} P_i$ over U_{ij} satisfying the cocycle condition $\phi_{ik} = \phi_{ij}\phi_{jk}$, one can glue the P_i to form a global object P of $\mathcal{C}_B$. We refer the reader to [**SGA1**] [**Bry**] [**Gi**] [**Bre**] for a detailed discussion. Then a gerbe $\mathcal{C}$ over B is a sheaf of categories $U \mapsto \mathcal{C}_U$ such that the following conditions are verified:

1) the category $\mathcal{C}_U$ is a groupoid, i.e. any morphism is invertible.

2) every point of B has a neighborhood U such that the category $\mathcal{C}_U$ is not empty;

3) any two objects of $\mathcal{C}_U$ are locally isomorphic.

In general, the sheaves of automorphisms $\underline{Aut}(P)$ of local objects of $\mathcal{C}$ are locally isomorphic to one another, but since there is no canonical such isomorphism, there may not exist a globally defined sheaf of groups G such that for an object P of $\mathcal{C}_U$, $\underline{Aut}(P)$ is isomorphic to $G_{/U}$. For our purposes, we will fix a sheaf of groups G such that for any local object P, the sheaf $\underline{Aut}(P)$ is locally isomorphic to G. We then say that the gerbe is of type G. We will also assume that $\underline{Aut}(P)$ is an inner form of G. This means that the transition cocycles take values in inner automorphisms of G. We will say that the gerbe is *inner*.

We say that a gerbe $\mathcal{C}$ is *neutral* if it has a global object. For instance, let $\underline{G}$ be any sheaf of groups. A $\underline{G}$-torsor is a sheaf of sets F, together with a left action of $\underline{G}$ on F such that F has local sections and given any $s \in F(U)$, the map

$\underline{G}(U) \to F(U), g \mapsto g \cdot s$, is an is a bijection. For instance, if $P \to M$ is a smooth principal bundle, then the sheaf F of smooth sections of $P \to M$ is a torsor under the sheaf $\underline{G}$ of smooth G-valued functions. The $\underline{G}$-torsors form a gerbe of type $\underline{G}$, which is neutral, as there is the trivial torsor.

Given ω, we let $\mathcal{C}_U$ be the category whose objects are quantum vector bundles (V, ∇) over U, and whose morphisms are connection-preserving isomorphisms. Conditions (1) and (2) are clear. For condition (3), consider two quantum vector bundles (V_1, ∇_1) and (V_2, ∇_2) of rank r over some open set U. Then the vector bundle $V_1^* \otimes V_2$ is equipped with a connection, and flat sections of this vector bundle correspond bijectively to connection preserving isomorphisms $V_1 \tilde{\to} V_2$. Now the curvature of the connection on $V_1^* \otimes V_2$ is $2\pi i(-r\omega + r\omega) = 0$, so there exists a flat section s which takes a given value at a given point x. If we pick this value to correspond to an invertible linear map $(V_1)_x \to (V_2)_x$, then for any other point y, the linear map $(V_1)_y \to (V_2)_y$ will also be invertible. Thus we have defined a gerbe. Clearly the automorphism of any object is locally isomorphic to the sheaf $GL(r, \mathbb{C})$ of locally constant $GL(r, \mathbb{C})$-valued functions.

Condition (1) of the theorem means that the gerbe $\mathcal{G}$ is neutral.

The gerbes of type $GL(r, \mathbb{C})$ are classified by the non-abelian cohomology set $H^2(M, GL(r, \mathbb{C}))$ of Giraud [**Gi**]. Let $Q_r \in H^2(M, GL(r, \mathbb{C}))$ be the class of the gerbe we just defined using quantum rank r bundles. There is a map $\alpha : H^2(M, \mathbb{C}^*) \to H^2(M, GL(r, \mathbb{C}))$ also due to Giraud. Here $H^2(M, \mathbb{C}^*)$ classifies gerbes of type $\mathbb{C}^*$, and is isomorphic to the usual Čech cohomology group. We have $\alpha(Q_1) = Q_r$. The class Q_1 is computed as follows:

LEMMA 2.3. *We have* $Q_1 = \exp(2\pi i[\omega]) \in H^2(M, \mathbb{C}^*)$.

This is proved in [**Bry**].

Now to finish the proof of the theorem we need only appeal to the exact sequence of sets established by Giraud [**Gi,** Chapter IV, Prop. 4.2.8]:

$$H^1(M, PGL(r, \mathbb{C})) \xrightarrow{\delta} H^2(M, \mathbb{C}^*) \xrightarrow{\alpha} H^2(M, GL(r, \mathbb{C})). \tag{2.4}$$

The meaning of this exact sequence is that for a class $c \in H^2(M, \mathbb{C}^*)$, the class $\alpha(c) \in H^2(M, GL(r, \mathbb{C})$ is neutral if and only if it is in the image of δ.

We briefly justify how neutral gerbes correspond to elements of $H^1(M, PGL(r, \mathbb{C}))$. As $PGL(r, \mathbb{C})$ is the group of inner automorphisms of $GL(r, \mathbb{C})$, the set $H^1(M, PGL(r, \mathbb{C}))$ classifies local systems of groups $H \to M$ which are locally isomorphic to $GL(r, \mathbb{C})$, and such that the transition cocycles are inner automorphisms. Such a local system of groups H determines a gerbe, whose objects over U are H-torsors over U; the gerbe is neutral, because the trivial H-torsor is a global object. Conversely, given a neutral gerbe $\mathcal{G}$, with global object P, the sheaf $\underline{Aut}(P)$ of automorphisms of P is locally isomorphic to $GL(r, \mathbb{C})$. As the gerbe is inner, this sheaf of groups corresponds to a class in $H^1(M, PGL(r, \mathbb{C}))$. It is then not hard to see that $\mathcal{G}$ is equivalent to the gerbe of $\underline{Aut}(P)$-torsors.

Let us illustrate the theorem in simple cases. In the case of a compact surface of genus ≥ 1, the r-torsion subgroup $[H^2(M, \mathbb{C}^*)]_r$ is isomorphic to $\mathbb{Z}/r$, and it is a well-known topological fact that the map $H^1(M, PGL(r, \mathbb{C})) \to \mathbb{Z}/r = [H^2(M, \mathbb{C}^*)]_r$ is surjective, Thus we recover Proposition 2.1. In contrast, if M is simply-connected, there exists a quantum vector bundle of rank r if and only ω is integral. In that case there is only one quantum vector bundle up to isomorphism, namely $L \oplus \cdots \oplus L$, where L is a quantum line bundle. Thus the existence of interesting higher rank quantum vector bundles is only possible in the non-simply connected case.

We note that in case there is quantum line bundle L then for any flat vector bundle V the tensor product bundle $L \otimes V$ is a quantum vector bundle, and conversely any quantum line bundle is obtained in this way.

3. Families of symplectic manifolds

In this section we consider the situation of a family $f : E \to B$ of symplectic manifolds. We study the problem of prequantizing the whole family of symplectic manifolds.

Here f is a smooth submersive mapping, and we have a fiberwise 2-form ω such that the restriction of ω to each fiber $E_x = f^{-1}(x)$ is closed and non-degenerate. In many interesting examples either B or the fibers of f are infinite-dimensional. If we assume that the restriction of ω to each fiber is quantizable, then it is easy to see that for a contractible open set $U \subset B$, there exists a line bundle over $f^{-1}(U) \subset E$, together with a fiberwise connection ∇ whose curvature is equal to $2\pi i\omega$. This is proved in [**Bry,** Chapter 5].

This leads to the notion of a quantum vector bundle in the context of the family $f : E \to B$ of symplectic manifolds. This will mean a vector bundle over E equipped with a fiberwise connection whose curvature is equal to $2\pi i\omega \otimes Id$. Again if there exists a quantum line bundle L, then for any vector bundle V equipped with a flat fiberwise connection, the tensor product $L \otimes V$ is a quantum vector bundle, and any quantum vector bundle is obtained in this way.

As in section 2, we can formulate the properties of the rank r quantum vector bundles in terms of a gerbe. For an open set $U \subseteq B$, we have the groupoid $\mathcal{G}_U$ whose objects are quantum vector bundles (V, ∇) of rank r over $\pi^{-1}(U)$, and whose morphisms are connection-preserving isomorphisms. If f is a smooth fibration with simply-connected fibers, then $\mathcal{G}$ is a gerbe on B of type $\underline{GL(r, \mathbb{C})}$, where $\underline{GL(r, \mathbb{C})}$ is the sheaf of smooth $GL(r, \mathbb{C})$-valued functions. In case $r = 1$, a gerbe of type $\underline{\mathbb{C}}^*$ is called in [**Bry**] a Dixmier-Douady gerbe. Dixmier-Douady gerbes are classified by an element of the Čech cohomology group $H^2(B, \underline{\mathbb{C}}^*)$, which is isomorphic to $H^3(B, \mathbb{Z})$ by the exponential exact sequence.

We explain how to associate a class in $H^2(B, \underline{\mathbb{C}}^*)$ to a Dixmier-Douady gerbe. We pick an open covering (U_i), objects P_i of $\mathcal{C}_{U_i}$, and isomorphisms $\phi_{ij} : P_j \to P_i$ over U_{ij}. Then we have

$$\phi_{ik} = c_{ijk}\phi_{ij}\phi_{jk}$$

for an automorphism c_{ijk} of P_i over U_{ijk}, which is a section of the sheaf $\underline{\mathbb{C}}^*$. Then c_{ijk} is a 1-cocycle with coefficients in $\underline{\mathbb{C}}^*$.

To find a closed 3-form representing this class, we developed in [**Bry,** Chapter 5] the differential geometry of Dixmier-Douady gerbes. We recall this next in our context. We wish to select a natural class of connections on a quantum vector bundle (E, ∇). We will look for connections $\tilde{\nabla}$ on E such that on the fibers of f, $\tilde{\nabla}$ restricts to ∇. To further narrow down the class of connections we consider, we need the following auxiliary data

(D) a 2-form β on E, which restricts to ω on the fibers of f, and such that $d\beta$ is the pull-back of a (closed) 3-form ν on B. Then we have

PROPOSITION 3.1 [**Bry**]. *The class of the gerbe $\mathcal{G}$ in $H^3(B, \mathbb{Z})$ is represented by the 3-form ν.*

When f is a smooth fibration with simply-connected fibers, we prove in [**Bry,** Lemma 4.3.3] that a 2-form β with these properties exists. An example is the evaluation map $f : LM \to M$, where M is a smooth manifold equipped with a closed 3-form ν with integer periods, LM is the free smooth loop space of a manifold M. Here $f(\gamma) = \gamma(0)$. There is then a natural 2-form β on LM such that

- the restriction of β to any fiber is closed and quantizable;
- $d\beta = \nu$.

In this context, there is no global quantum line bundle unless ν is exact. Again, as in section 2, this can be formulated in terms of a suitable gerbe.

Another example starts with a compact manifold M which admits a symplectic form ω with integral cohomology class. Then we can take B to the set of all 1-forms α on M such that $\omega + d\alpha$ is symplectic. This is an open set in the Fréchet space of all 1-forms, so it is a Fréchet manifold. We set $E = B \times M$, and let $f = p_1 : B \times M \to B$ be the first projection. There is a tautological 1-form η on $B \times M$, and $\beta = \omega + d\eta$ satisfies our conditions (with $\nu = 0$ since β is closed).

Given the 2-form β satisfying (D) we can narrow down the class of connections on a quantum vector bundle (V, ∇) to the following:

(i) $\tilde{\nabla}$ is a connection on V, which restricts to ∇ on the fibers;

(ii) if K denotes the curvature of ∇, we have:

$$K(v, w) - 2\pi i \beta(v, w)\, Id = 0 \tag{3.1}$$

for tangent vectors v, w to E with v vertical.

In the rank one case we have

LEMMA 3.2. *Assume V is a line bundle. Then equation (3.1) implies that $K - 2\pi i$ is the pull-back of a complex-valued 2-form on B.*

In this case by Lemma 3.2 we obtain a complex 2-form, say F, on B. We take the position that the 2-form F is really the relevant curvature form attached to (V, ∇) since it lives on B, as opposed to K itself which lives on the total space E.

We call F the descended curvature form. We note the following:

$$dF = 2\pi i \nu \tag{3.2}$$

Now we consider the higher rank case $r > 1$. Then $R = K - 2\pi i\omega \otimes id$ satisfies the following twisted version of the Bianchi identity:

$$\nabla R = 2\pi i \nu \otimes Id \tag{3.3}$$

We will construct a matrix-valued 2-form on B which will be the descended curvature. We decompose the bundle $End(V)$ into the direct sum

$$End(V) = \mathbb{C} \oplus End_0(V),$$

where $\mathbb{C}$ consists of the scalar matrices, and $End_0(V)$ is comprised of the matrices of trace zero. The curvature K decomposes accordingly as

$$K = K_{sc} + K_0 \tag{3.4}$$

We will construct separately the two components of the 2-form. For the scalar part, we consider the 2-form $\frac{1}{r}\, Tr(K) - 2\pi i\omega$; just as in Lemma 2.2, this descends to a scalar-valued 2-form F_{sc}. For the non-scalar component, we first construct a Lie algebra bundle $A \to B$ with fiber $\mathfrak{sl}(r, \mathbb{C})$. We start with the Lie algebra bundle

$End_0(V)$ over E, which is associated with the representation $\mathfrak{sl}(r,\mathbb{C})$ of $GL(r,\mathbb{C})$. We have a connection on $End_0(V)$ induced from $\tilde{\nabla}$; from equation (3.1) we see that this connection has zero curvature on each fiber of f. If we now assume

(iii) the fibers of f are simply-connected

then the flat fiberwise connection allows us to descend $End_0(V)$ to a Lie algebra bundle $A \to E$. We claim that the $End_0(V)$-valued 2-form K_0 on E descends to an A-valued 2-form. First (ii) says that K_0 is purely horizontal. Then we claim that for two vector fields v, w on B and for any lifts $\tilde{v}, \tilde{w}$ to vector fields on E, the A-valued function $x \mapsto K_o(\tilde{v}, \tilde{w})$ is fiberwise killed by the connection ∇. This can be checked locally, in a local trivialization of A inducing a local trivialization of $End_0(V)$. In that context, if α is the potential of the connection $\tilde{\nabla}$, it follows from the Bianchi identity $dK_0 = [\alpha, K_0]$ that we have, for any vertical vector field ξ:

$$\xi \cdot K_0(\tilde{v}, \tilde{w}) = (\mathcal{L}_\xi K_0)(\tilde{v}, \tilde{w}) = [\alpha(\xi), K_0(\tilde{v}, \tilde{w})],$$

which means that $K_0(v, w)$ is a horizontal section of A along the fibers of f. Hence K_0 descends to an A-valued 2-form on B.

This means that K_0 is a basic 2-form with values in A, as we claimed.

We then define the bundle of Lie algebras $\tilde{A} = \mathbb{C} \oplus A$ over B, with typical fiber $M_r(\mathbb{C})$. The pull-back of $\tilde{A}$ to E identifies with $End(V)$. The descended curvature of $\tilde{\nabla}$ is defined as the $\tilde{A}$-valued 2-form

$$F = \left[\tfrac{1}{r}\, Tr(K) - 2\pi i \omega\right] \oplus K_0 \tag{3.5}$$

We want to view F as a type of field on B which incorporates part of the geometry of the family of symplectic manifolds parameterized by B. As usual it is important to understand what is the space of all admissible connections $\tilde{\nabla}$. We have

PROPOSITION 3.3. *If (V, ∇) is a quantum line bundle over E, then the space of admissible connections over (V, ∇) is an affine space under the space $A^1(B)$ of complex-valued 1-forms on B.*

In the terminology of [**Bry**], this means that our notion of admissible connections gives a connective structure on the gerbe of Proposition 3.1.

If B is an oriented riemannian four-dimensional manifold we can consider the usual Yang-Mills action

$$S(\nabla) = \int_M Tr(F \wedge *F) \tag{3.6}$$

Of course this makes sense precisely because the descended curvature F is a 2-form on the base B. We wish to write down the equations corresponding to the extremal fields for the functional S. This is at least not hard to do in the rank one case. The space of admissible connections is described as an affine space in Proposition 3.3. We can vary S with respect to the vector space $A^1(B)$ and we see that $\tilde{\nabla}$ is a critical point of S if and only if ∇ satisfies the equation $d * F = 0$. Since we have $dF = \Omega = 2\pi i \nu$ by construction, we obtain the system of equations

$$dF = \Omega, \quad d * F = 0 \tag{3.7}$$

which may be viewed as a version of the Maxwell equations in presence of matter fields.

We note the following

PROPOSITION 3.4. *For a compact oriented riemannian four-manifold B and for any exact 3-form ν on B, the twisted Maxwell equations (3.7) have a smooth global solution F.*

PROOF. Let $A^j_{(2)}(B)$ denote the Hilbert space of L_2 j-forms on B. We have the Hodge decomposition:

$$A^2_{(2)}(B) = Ker(d) \oplus Im(\delta).$$

Write $\Omega = d\beta$ and write $\beta = \beta_1 + \delta\alpha$ where $d\beta_1 = 0$. It is well-known that α is smooth. Then $\Omega = d\delta\alpha$, and $F = \delta\alpha$ is a solution of (3.7). □

In the higher rank case we do not know precisely the structure of the space of admissible connections. However we still have the following weaker analog of Proposition 3.3. We use the algebra bundle $\tilde{A}$ over B.

PROPOSITION 3.5. *If $\tilde{\nabla}$ is an admissible connection and if α is an $\tilde{A}$-valued 1-form on B, then the connection $\tilde{\nabla} + f^*\alpha$ is still admissible.*

PROOF. This follows since the derivative of the curvature in the direction of the $\tilde{A}$-valued 1-form α is equal to the skew-symmetric part of the covariant derivative $\tilde{\nabla}\alpha$, and the contraction of this 2-form with any vertical vector field vanishes. □

It follows from Proposition 3.5 that we can use variational calculus with respect to the space of $\tilde{A}$-valued 1-forms.

Thus we find, just as for the ordinary Yang-Mills functional, that the extremal points of S on the space of admissible connections must satisfy the Yang-Mills equation

$$\tilde{\nabla} * F = 0. \tag{3.8}$$

The new aspect here is that the space of ordinary connections is replaced with the space of admissible connections. Their curvature verifies the twisted Bianchi identity (3.3).

It is not clear however if this equation is sufficient to ensure that $\tilde{\nabla}$ is an extremal point for S.

Let (V, ∇) be a quantum vector bundle over E. There is an easy way to construct other quantum vector bundles over E by tensor product. Let H be some auxiliary a vector bundle over B. Then the pull-back bundle f^*H has a fiberwise connection. There is a fiberwise connection on the tensor product bundle $V \otimes f^*H$. Clearly this is a quantum vector bundle over E. Now consider an admissible connection $\tilde{\nabla}$, i.e., satisfies (i)-(ii). Pick some connection D on f^*H which agrees with the given fiberwise connection. It is then clear that the tensor product connection on the tensor product bundle $V \otimes f^*H$ is again admissible.

We summarize this by saying that our class of vector bundles and admissible connections is stable under tensor product with vector bundles with connections over the base B.

4. Projective space bundles

Let B be a smooth manifold (finite or infinite-dimensional), and let $f : \mathbb{P} \to B$ be a bundle of projective spaces (finite or infinite-dimensional). We assume that the typical fiber is the projective space $\mathbb{P}(E)$ of lines in a complex Hilbert space

E, and that the structure group of the fibration is the group $PGL(E)$ of projective automorphisms of $\mathbb{P}(E)$. The group $PGL(E)$ in turn is the quotient of the group $GL(E)$ of bicontinuous linear automorphisms of E by the subgroup $\mathbb{C}^*$ of scalar transformations. $PGL(E)$ is a Lie group in the sense of [**P-S**].

We have then the notion of a vector bundle $\mathcal{H}$ over B whose associated projective space bundle is isomorphic to $f : \mathbb{P} \to B$. The data of such a vector bundle is equivalent to a lifting of the structure group of f to the group $GL(E)$. Accordingly such a vector bundle $\mathcal{H} \to B$ will be called a vector bundle over B *lifting* $f : \mathbb{P} \to B$, or simply a vector bundle lift of $f : \mathbb{P} \to E$. Such liftings exist at least locally on B, for instance on any open set over which the $PGL(H)$-bundle is trivial. If $\mathcal{H}$ is a vector bundle over B lifting $f : \mathbb{P} \to E$ and if $L \to B$ is any line bundle, then the tensor product $\mathcal{H} \otimes L$ is another vector bundle lift of $f : \mathbb{P} \to B$. Conversely, any other vector bundle lift $\mathcal{H}'$ is of the form $\mathcal{H} \otimes L$ for a unique line bundle L over B. We can describe the corresponding $\mathbb{C}^*$-bundle $L^* \to B$ as follows. The sections of L^* over some open set $U \subset B$ are those vector bundle isomorphisms $\phi : \mathcal{H} \tilde{\to} \mathcal{H}'$ over U such that the following diagram commutes

$$\begin{array}{ccc} \mathbb{P}(\mathcal{H}) & \xrightarrow{\phi} & \mathbb{P}(\mathcal{H}') \\ \downarrow & & \downarrow \\ \mathbb{P} & \xrightarrow{Id} & \mathbb{P} \end{array} \tag{4.1}$$

All this can be formulated in terms of gerbes: there is a gerbe $\mathcal{G}$ over B whose objects are vector bundles $\mathcal{H} \to B$ lifting $\mathbb{P}_{/U} \to U$. As the automorphism sheaf of such a lifting is equal to $\underline{\mathbb{C}}^*$, the gerbe $\mathcal{G}$ is a Dixmier-Douady gerbe. It is a general property of Dixmier-Douady gerbes that given any object P over some open set. any other object is obtained by twisting P by a smooth $\mathbb{C}^*$-bundle.

We now wish to consider more generally vector bundles over B which are of the form $V = \mathcal{H} \otimes W$, where $\mathcal{H}$ is a vector bundle lifting $f : \mathbb{P} \to B$, and W is any vector bundle (of finite rank) over B. We say that such $V \to B$ is a vector bundle of type $\mathbb{P} \to B$ over B. We have the following lemma

LEMMA 4.1. *(1) The direct sum of two vector bundles of type $\mathbb{P} \to B$ is another vector bundle of type $\mathbb{P} \to B$.*

(2) If $V \to B$ is a vector bundle of type $\mathbb{P} \to B$ and $W \to B$ is a finite rank vector bundle, the tensor product $V \otimes W$ is again of type $\mathbb{P} \to B$.

PROOF. Let $V_1 = \mathcal{H}_1 \otimes W_1$ and $V_2 = \mathcal{H}_2 \otimes W_2$ be two vector bundles of type $\mathbb{P} \to E$, where $\mathcal{H}_j \to B$ is a vector bundle lifting f. Since $\mathcal{H}_2 = \mathcal{H}_1 \otimes L$ for a line bundle L, we may assume that $\mathcal{H}_2 = \mathcal{H}_1$, in which case (1) is obvious. (2) is immediate. □

We need to introduce a finite-dimensional algebra bundle A over B, with typical fiber $M_r(\mathbb{C})$, attached to a vector bundle V of type f over B. If we write $V = \mathcal{H} \otimes W$, where $\mathcal{H}$ is a lift of f and f has finite rank r, then we define A as the algebra bundle $End(W)$. Now let $\mathcal{H}'$ be another vector bundle lift of f, so that by Lemma 4.1 $\mathcal{H}' = \mathcal{H} \otimes L$ for a unique line bundle L.Then we have $V = \mathcal{H}' \otimes (V \otimes L^{-1})$, and of course $End(V \otimes L^{-1}) \tilde{\to} End(V)$. We call A the small algebra bundle attached to V.

The construction of the small algebra bundle can be phrased in the language of principal bundles. A vector bundle V of type f corresponds to a principal $GL(E \otimes \mathbb{C}^r)$-bundle with the following properties:

(a) the structure group is restricted to the product group $[GL(E) \times GL(r, \mathbb{C})]/S$, where $S = \mathbb{C}^*$ embedded in $GL(E) \times GL(r, \mathbb{C})$ by $z \in \mathbb{C}^* \mapsto (z, z^{-1})$.

(b) the corresponding $PGL(E)$-bundle over B is identified with $f : \mathbb{P} \to B$.

In (b) we use the natural group homomorphism $[GL(E) \times GL(r, \mathbb{C})]/S \to PGL(E)$ (first projection).

Then using the group homomorphism $[GL(E) \times GL(r, \mathbb{C})]/S \to PGL(r, \mathbb{C})$ (second projection) we obtain a $PGL(r, \mathbb{C})$-bundle over B. Since $PGL(r, \mathbb{C})$ is the group of automorphisms of the algebra $M_r(\mathbb{C})$, this yields an algebra bundle over B with fiber $M_r(\mathbb{C})$. This is nothing else but the small algebra bundle $A \to B$.

We now turn to the notion of admissible connections for vector bundles of type $\mathbb{P} \to B$. First we fix a projective connection θ on the projective space bundle $f : \mathbb{P} \to B$. This means we have a connection on the corresponding principal $PGL(E)$-bundle. Then if we have a vector bundle $\mathcal{H} \to B$ lifting f, we can consider all the connections $\tilde{\theta}$ on the principal $GL(E)$-bundle which lift θ. Such a connection $\tilde{\theta}$ will be admissible. We have

LEMMA 4.2 [**Bry**]. *Given a connection θ on $f : \mathbb{P} \to B$ and a vector bundle $\mathcal{H} \to B$ lifting f, the space of admissible connections $\tilde{\theta}$ on $\mathcal{H} \to B$ is an affine space under $A^1(B)$.*

Now we extend this notion to an arbitrary vector bundle of type $f : \mathbb{P} \to B$, say $V = \mathcal{H} \otimes W$. We say that a connection on V is admissible if it is of the form $\nabla = \tilde{\theta} \otimes 1 + 1 \otimes D$, where $\tilde{\theta}$ is an admissible connection on $\mathcal{H}$ and D is a connection on $W \to B$. It is easy to see that this condition is independent of the choice of the lifting $\mathcal{H}$ involved in expressing V as a tensor product $V \otimes W$. Then we have the following analog of Lemma 3.2.

LEMMA 4.3. *Let V be a vector bundle over B of type $f : \mathbb{P} \to B$, and let A be the corresponding small algebra bundle over B. Then the set of admissible connections over V is a an affine space under the vector space of A-valued 1-forms over B.*

Now just as in section 2 we wish to introduce the notion of curvature of an admissible connection, which should be a 2-form on B with values in A. This requires some extra piece of data, as the curvature K of an admissible connection $\tilde{\theta} + D$ on $V = \mathcal{H} \otimes W$ is of the form $K = R_1 \otimes Id + Id \otimes R_2$, where

- the curvature R_1 of $\tilde{\theta}$ is a 2-form with values in $End(\mathcal{H})$

- the curvature R_2 of D is a 2-form with values in $A = End(W)$.

So $R_1 + \otimes Id + Id \otimes R_2$ is a 2-form with values in the algebra bundle $End(\mathcal{H}) \otimes End(W) = End(V)$. In fact it takes values in the smaller algebra $[End(\mathcal{H}) \oplus End(W)]/\mathfrak{s}$, where $\mathfrak{s} = \mathbb{C}$ embedded into $End(\mathcal{H}) \oplus End(W)$ by $z \mapsto (z, -z)$. So it is clear that the piece of data needed is

(c) a linear form $l : End(\mathcal{H}) \to \mathbb{C}$ such that $l(Id) = 1$.

Of course there are many such linear forms, for instance $l(T) = \langle Tv, v \rangle$ for some unit vector v. We have a corresponding linear map

$$l \otimes Id : End(\mathcal{H}) \otimes End(W) \to End(W). \tag{4.2}$$

Then we define the *descended curvature* of $\tilde{\theta}+D$ to be the $A = End(W)$-valued 2-form

$$F = (l \otimes Id)(R_1 \otimes Id + Id \otimes R_2) \tag{4.3}$$

Just as in §3 the descended curvature F satisfies a twisted Bianchi identity

$$DF = \Omega \otimes Id \tag{4.4}$$

where Ω is a complex-valued closed 3-form on B, called the 3-curvature. To describe the 3-form Ω locally on B, we can assume that the bundle $\mathbb{P} \to B$ is trivial over some open set $U \subset B$. Then the given connection θ is given by a 1-form A on U with values in $End(E)/\mathbb{C}$. We can lift this 1-form to an $End(E)$-valued 1-form $\tilde{A}$. Then we can form the curvature form $R = d\tilde{A} + \frac{1}{2}[\tilde{A}, \tilde{A}]$. Using the linear map $l : End(E) \to \mathbb{C}$ we obtain a scalar-valued 2-form $\beta = l(R)$. We then have

$$\Omega = d\beta. \tag{4.5}$$

Note that if we change $\tilde{A}$ to $\tilde{A} + \alpha$ where α is a complex-valued 1-form, we change R to $R + d\alpha \otimes Id$, hence we change β to $\beta + d\alpha$. So the 3-form Ω does not depend on the choice of $\tilde{A}$; therefore it is globally defined on B.

The following result is proved in [**Bry**]

PROPOSITION 4.4. *The cohomology class of* $\frac{\Omega}{2\pi i}$ *is integral. Its class is the image in* $H^3(B, \mathbb{Z})$ *of the class of the Dixmier-Douady gerbe of liftings of* $\mathbb{P} \to B$.

In other words, the cohomology class of Ω is the obstruction to finding a global lifting $\mathcal{H} \to B$ of the projective space bundle $\mathbb{P} \to B$.

We refer to [**Bry**] for the detailed study of an example where $B = S^2 \times S^1$, and the fiber over (x, z) of the projective bundle is the projective space associated to a vector space of multi-valued functions (with monodromy) over the universal covering space of the fiber over x of the Hopf fibration. In that case the structure group of the bundle is reduced to the Heisenberg group associated to $\mathbb{Z}$ viewed as a locally compact abelian group.

If we now assume that B is a closed oriented riemannian 4-manifold we can then introduce the Yang-Mills functional

$$YM(\nabla) = \int_M Tr(F \wedge *F) \tag{4.6}$$

We then have the following characterization of the extremal points of the Yang-Mills functional:

PROPOSITION 4.5. *The admissible connection* $\nabla = \tilde{\theta} + D$ *is an extremal point for the functional* YM *if and only if we have*

$$D * F = 0.$$

PROOF. This follows form the usual calculus of variations, using the description of the space of connections given in Lemma 4.3. □

There is a somewhat more general context in which the constructions of this section can be carried out. The starting point would then be a principal G-bundle

$f : E \to B$ where G is a Lie group (maybe infinite-dimensional). As auxiliary data we need a central extension of Lie groups

$$1 \to \mathbb{C}^* \to \tilde{G} \to G \to 1.$$

One also needs some representation E of $\tilde{G}$. This leads one to consider vector bundles over B which are of the type $\tilde{E} \otimes W$, where

- $\tilde{E}$ is locally isomorphic to a vector bundle associated with a $\tilde{G}$-bundle lifting f and the representation E of $\tilde{G}$.
- W is a finite dimensional vector bundle.

The notion of admissible connection and of descended curvature is then introduced in the same way as before.

5. 2-vector spaces, 2-vector bundles and gerbes

In the two types of situations analyzed in §3 and §4, we have the following type of geometric data:

1) some class of vector bundles over a base manifold B;
2) a natural closed 3-form Ω on B.

The cohomology class of $\nu = \dfrac{\Omega}{2\pi i}$ is integral and represents an obstruction class, namely the obstruction to the existence of some global vector bundle in the class. This is phrased in terms of gerbes and their differential geometry. There is another categorical language to study this type of situation, namely the 2-vector spaces of Kapranov and Voevodsky [**K-V**].

First of all we focus on the simple case where B is just a point. Of course in that case there is no 3-form, but the underlying categorical structure will be apparent. Turning our attention to the first example studied in §2, we consider a symplectic manifold (M, ω). Then we have introduced the notion of a quantum vector bundle (E, ∇). The collection of these quantum vector bundles has the structure of category $\mathcal{C}$. Given quantum vector bundles (E_1, ∇_1) and (E_2, ∇_2), a morphism from (E_1, ∇_1) to (E_2, ∇_2) is a vector bundle map $\phi : E_1 \to E_2$ which is compatible with the connections. This implies that the rank of ϕ is constant, so that we have an exact sequence of vector bundles

$$0 \to Ker(\phi) \to E_1 \to E_2 \to Coker(\phi) \to 0 \tag{5.1}$$

and $Ker(\phi)$, $Coker(\phi)$ are quantum vector bundles. Therefore $\mathcal{C}$ has the structure of an abelian category. We also have another operation, namely if (V, ∇) is a quantum vector bundle and (W, D) is a vector bundle equipped with a flat connection, then the tensor product $(V \otimes W, \nabla)$ is another quantum vector bundle. If we assume M simply-connected, then we can think W as the constant flat bundle associated to a vector space W.

The structure of the category $\mathcal{C}$ can be summarized using the concept of 2-vector space.

Definition 5.1 (Kapranov-Voevodsky [**K-V**]). A 2-vector space (of 2-rank one) is an abelian category $\mathcal{C}$ equipped with the operation of tensor product of objects of $\mathcal{C}$ with finite-dimensional vector spaces, such that there is a generator L of $\mathcal{C}$, that is to say an object L such that the functor $W \mapsto L \otimes W$ is an equivalence of categories from the category of vector spaces to $\mathcal{C}$.

We refer to [**K-V**] for a detailed discussion of 2-vector spaces and for examples. In this paper we only consider 2-vector spaces of 2-rank one. This notion of 2-rank should not be confused with the notion of rank of an object P of $\mathcal{C}$: P is of rank r if $P = L \otimes W$, where L is a generator of $\mathcal{C}$ and W is a vector space of dimension r; equivalently, the isomorphism group of P is $GL(r, \mathbb{C})$.

Then we have

PROPOSITION 5.2. *The category of quantum vector bundles over a quantizable simply-connected symplectic manifold* (M, ω) *has the structure of a* 2*-vector space.*

Now we turn to the more general situation of a family $f : E \to B$ of symplectic manifolds over B. We introduced in §3 the notion of quantum vector bundle (V, ∇) over E. We can then form a category $\mathcal{C}_B$ whose objects are quantum vector bundles (V, ∇) over E. We take the morphisms in $\mathcal{C}_B$ to be vector bundle maps which are compatible with the fiberwise connections. Compared with the case B is a point, the new aspect here is that such a homomorphism need not have constant rank, so its kernel and cokernel will not be objects of $\mathcal{C}_B$. Therefore $\mathcal{C}_B$ is an additive category but not an abelian category.

We assume for now that the fibers of f are simply-connected. Then assuming the category $\mathcal{C}_B$ non-empty, we find the following structure on it:

1) it is an additive category

2) we have the operation of tensoring an object of $\mathcal{C}_B$ with a vector bundle over B.

Now we would like to state that there is a generator of $\mathcal{C}_B$, say V, such that any other object is of the form $V \otimes W$ for some vector bundle W. This is not quite true in general, because there may not be a global quantum line bundle. However it will be true "locally on B". To formalize this idea, we introduce the category $\mathcal{C}_U$ for any open set $U \subset B$ and note that we have a restriction functor $\mathcal{C}_U \to \mathcal{C}_V$ whenever $V \subset U$. Then we have

3) for any $x \in B$ there exists an open neighborhood U of x such that the category $\mathcal{C}_U$ has a free generator P. This means that $W \mapsto P \otimes W$ is an equivalence from the category of vector bundles to $\mathcal{C}_U$.

We express 3) by saying that $\mathcal{C}_B$ has a generator locally.

We also have:

4) The assignment $U \mapsto \mathcal{C}_U$ is a sheaf of categories (i.e., a stack) over B.

This leads us to the notion of 2-vector bundle over B:

DEFINITION 5.3. A 2-vector bundle (of 2-rank one) over B consists of the data for each open set $U \subset B$ of an additive category $\mathcal{C}_U$ together with restriction functors $\mathcal{C}_U \to \mathcal{C}_V$ when $V \subset U$ and the operation of tensor product $V \otimes W$ of an object V of $\mathcal{C}$ with a vector bundle W, such that properties 1)-4) are verified.

So we can state

PROPOSITION 5.4. *If* $(E \to B, \omega)$ *is a quantizable family of symplectic manifolds with simply-connected fibres, then the assignment* $(U \hookrightarrow B) \mapsto \mathcal{C}_U$ *is a* 2*-vector bundle over* B.

In §4 we studied the geometry of vector bundles of type $f : \mathbb{P} \to B$, where $f : \mathbb{P} \to B$ is a projective space bundle. Recall such a vector bundle was of the form $\mathcal{H} \otimes W$, where $\mathcal{H} \to B$ is a vector bundle over B lifting $\mathbb{P} \to B$, and W is a finite rank vector bundle. This leads to a 2-vector bundle over B as follows. If U is an open set in B the vector bundles over U of type f form a category. We describe the set of morphisms from $\mathcal{H} \otimes W_1$ to $\mathcal{H} \otimes W_2$:

$$Hom(\mathcal{H} \otimes W_1, \mathcal{H} \otimes W_2) = Hom(W_1, W_2) \tag{5.2}$$

in which $Hom(W_1, W_2)$ is the set of vector bundle maps $W_1 \to W_2$. It is easily seen that this description is independent of the choice of the vector bundle $\mathcal{H}$ lifting $f_{/U}$. We then have without difficulty:

PROPOSITION 5.5. *The assignment $U \mapsto \mathcal{C}_U$ yields a 2-vector bundle intrinsically attached to the projective space bundle $f : \mathbb{P} \to B$.*

Thus both geometric situations discussed in §3 and §4 lead to a 2-vector bundle. On the other hand, we previously interpreted them in terms of gerbes. In each case there is a natural notion of rank of an object of $\mathcal{C}_U$. The notion of "rank" requires some explanation in the case of a projective space bundle $\mathbb{P} \to B$. In this case we define the rank of $\mathcal{H} \otimes W$ to be the rank of W. So in both cases an object of rank 1 in $\mathcal{C}_U$ gives a generator of that category. Equivalently, an object P of $\mathcal{C}_U$ is of rank r if the sheaf $\underline{Aut}(P)$ of automorphisms of P is locally isomorphic to $\underline{GL(r, \mathbb{C})}$. Thus inside sheaf of categories $\mathcal{C}$ we have the sheaf of groupoids $\mathcal{C}_r$ comprised of those objects which have rank r. This sheaf of categories $\mathcal{C}_r$ is a gerbe of type $\underline{GL(r, \mathbb{C})}$.

More generally the relation between 2-vector bundles and gerbes is summarized in the following:

PROPOSITION 5.6. *Any 2-vector bundle (of 2-rank one) over B gives rise to a Dixmier-Douady gerbe over B. Conversely any Dixmier-Douady gerbe can be extended to a 2-vector bundle over B.*

We briefly explain how to go from a Dixmier-Douady gerbe $\mathcal{G}$ to a 2-vector bundle $\mathcal{C}$. We observe that if P is any object of $\mathcal{G}_U$, then for any line bundle L over U we can construct we can construct a new object $P \otimes L$ of $\mathcal{G}_U$. To do this, cover U by open sets V_i and let g_{ij} be a Čech cocycle for a system of non-vanishing sections s_i of L over V_i. Then each g_{ij} defines an automorphism of $\mathcal{G}$ over V_{ij}. Since they satisfy the cocycle relation, we can use them to construct a new object of $\mathcal{G}$ over U_i. This object is independent, up to a canonical isomorphism, of the choice of the open covering and of the sections s_i. It will be denoted by $P \otimes L$. Conversely, given any other object Q over $\mathcal{U}$, the sheaf $Isom(P, Q)$ is a $\underline{\mathbb{C}}^*$-torsor, hence is the sheaf of sections of a principal $\mathbb{C}^*$-bundle $R \to U$. For L the corresponding line bundle, Q identifies with $P \otimes L$. Thus the construction $L \mapsto P \otimes L$ establishes an equivalence between the gerbe of all line bundles and the gerbe $\mathcal{G}$ over U.

Now given the Dixmier-Douady gerbe $\mathcal{G}$ over B, we will construct a 2-vector bundle $\mathcal{C}$. Roughly speaking, objects of $\mathcal{C}$ look locally like $P \otimes E$, where P is an object of $\mathcal{G}$ and E is a vector bundle. So we first construct a presheaf of categories $\mathcal{D}$ over B, whose objects over U are tensor products $P \otimes E$, where P is an object of $\mathcal{C}_U$ and E is a vector bundle over U. If P' is another object of $\mathcal{G}_U$, then Q is $P \otimes L$ for as line bundle L. Then if V' is a vector bundle over U, we put

$$Hom_{\mathcal{D}}(P \otimes E, P' \otimes E') = Hom_U(E, L \otimes E').$$

Then we construct the sheaf of categories $\mathcal{C}$ by sheafifying $\mathcal{D}$. Thus an object of $\mathcal{C}_U$ is obtained from an open covering (V_i) of U, and objects $P_i \otimes E_i$ of $\mathcal{D}$ over V_i, together with isomorphisms

$$(P_i \otimes E_i)_{V_{ij}} \tilde{\to} (P_j \otimes E_j)_{V_{ij}}$$

which satisfy the cocycle relation. We can tensor an object of $\mathcal{C}_U$ with a vector bundle in the natural way, and it is easy to see that $\mathcal{C}$ is a 2-vector bundle of 2-rank one.

Even the case where B is a point is interesting.. We get in that case the correspondence between the notion of gerbe over a point and the notion of a 2-vector space. In that case there is of course no cohomological obstruction to finding a global object. However if we have a group G acting on the gerbe we can define as in [**Bry,** Chapter 7] the notion of an equivariant object of the gerbe. The obstruction to constructing such an equivariant object is a class in $H^3(G, \mathbb{C}^*)$. The same therefore applies to the question of finding an equivariant object of rank 1 in a 2-vector space over which G operates. Presumably the question of finding a rank r object on which G operates leads to a class in a non-commutative cohomology group $H^3(G, GL(r, \mathbb{C}))$.

We now turn to the question of finding a global object of rank r in $\mathcal{C}_B$. We have the gerbe $\mathcal{C}_r$ whose objects are the rank r objects of $\mathcal{C}$. Let $\kappa \in H^3(B, \mathbb{Z})$ be the class attached to the Dixmier-Douady gerbe $\mathcal{C}_1$. We explain how the gerbe $\mathcal{C}_r$ can be described by a non-commutative Čech 2-cocycle with values in $\underline{GL(r, \mathbb{C})}_B$.

Pick a good open covering (U_i), objects P_i of the gerbe $\mathcal{C}_r$ over U_i and isomorphisms $\phi_{ij} : P_j \tilde{\to} P_i$ over U_{ij}. Then one has

$$\phi_{ik} = c_{ijk}\phi_{ij}\phi_{jk}$$

for a unique $c_{ijk} \in Aut((P_i)_{/U_{ijk}})$. The pair (c_{ijk}, ϕ_{ij}) is called a non-abelian 2-cocycle. The following cocycle conditions are verified:

$$\phi_{ik} = \phi_{ij}\phi_{jk}, c_{ikl}c_{ijk} = c_{ijl}[\overline{\phi_{ij}}c_{jkl}], \tag{5.3}$$

where $\overline{\phi_{ij}} : Aut((P_j)_{/U_{ij}}) \tilde{\to} Aut((P_i)_{/U_{ij}})$ is the isomorphism induced by ϕ_{ij}. If we fix an isomorphism ψ_i between $\underline{Aut}(P_i)$ and $\underline{GL(r, \mathbb{C})}_{U_i}$, then we can view ϕ_{ij} as an automorphism of $GL(r, \mathbb{C})$ and c_{ijk} as a map $U_{ijk} \to GL(r, \mathbb{C})$. Since we are working with "inner" gerbes, there is a natural choice of ψ_i modulo inner automorphisms, and then each ϕ_{ij} is an inner automorphism.

Returning to the class in $H^2(B, \underline{GL(r, \mathbb{C})}_B)$ corresponding to $\mathcal{C}_r$, it is easy to construct a non-abelian cocycle representing it. If γ_{ijk} is the Čech 2-cocycle with coefficients in $\underline{\mathbb{C}}^*_B$ representing the gerbe $\mathcal{G} = \mathcal{C}_1$, then we have the cocycle $\gamma_{ijk}\, Id$ with coefficients in $\underline{GL(r, \mathbb{C})}$.

PROPOSITION 5.7. *If there exists a global object of* $\mathcal{C}$ *of rank* r, *then* $r\kappa = 0$ *in* $H^3(B, \mathbb{Z})$.

PROOF. If a gerbe of type $\underline{GL(r, \mathbb{C})}_B$ is neutral, the class of the gerbe must be in the kernel of the so-called abelianization map (in the sense of Borovoi [**Bor**])

$$Ab : H^2(B, \underline{GL(r, \mathbb{C})}_B) \to H^2(B, \underline{\mathbb{C}}^*_B) \tag{5.4}$$

This map Ab is simply a determinant map on the level of non-abelian 2-cocycles.

The abelianization map sends the class of the cocycle (ϕ_{ij}, c_{ijk}) to the class of the cocycle $det(c_{ijk})$. Thus, in our example, the abelianization map sends $\gamma_{ijk}\, Id$ to its r-th power γ^r_{ijk}. The existence of a global object of $\mathcal{D}_U$ implies that the class of γ^r_{ijk} in $H^2(B, \underline{\mathbb{C}}^*_B) = H^3(B, \mathbb{Z})$ is trivial. □

We will give a more precise result. For this we will use (as in §2) the coboundary map $\delta : H^1(B, \underline{PGL(r, \mathbb{C})}) \to H^2(B, \underline{\mathbb{C}}^*) = H^3(B, \mathbb{Z})$, whose image consists of classes killed by r. Then we have

PROPOSITION 5.8. *There exists a global object of $\mathcal{C}$ of rank r if and only if the class $\kappa \in H^2(M, \underline{\mathbb{C}}^*) = H^3(B, \mathbb{Z})$ belongs to the image of*

$$\delta : H^1(B, \underline{PGL(r, \mathbb{C})}) \to H^2(B, \underline{\mathbb{C}}^*) = H^3(B, \mathbb{Z}).$$

The proof is entirely similar to that of Theorem 2.2.

The correspondence between gerbes and 2-vector spaces can be extended to the differential geometric structures on these objects. For gerbes we introduced in [**Bry**] the notions of connective structure and curving. There are corresponding notions for 2-vector bundles: the admissible connections developed in §2 and §3 fit in this context.

It is tempting to wonder about the possibility of essentially non-abelian extensions of the notions of 2-vector spaces and 2-vector bundles, and whether they might lead to interesting non-abelian cohomology.

References

[A-B] M. F. Atiyah and R. Bott, *The Yang-Mills equations over Riemann surfaces*, Phil. Trans. Royal Soc. London **308** (1982), 523–615.

[Bor] M. V. Borovoi, *Abelianization of the second non-abelian Galois cohomology*, Duke Math. J. **72** (1993), 217–239.

[Bre] L. Breen, *Bitorseurs et cohomologie non-abélienne*, The Grothendieck Festschrift I (P. Cartier et al., eds.), Progress in Math., vol. 86, Birkhäuser, Boston, 1990, pp. 401–476.

[Bry] J.-L. Brylinski, *Loop Spaces, characteristic classes and geometric quantization*, Progress in Math., vol. 107, Birkhäuser, Boston, 1993.

[Bry-ML] J-L. Brylinski and D. McLaughlin, *The geometry of the first Pontryagin class and of line bundles on loop spaces I*, Duke Math. Jour. **35** (1994), 603-638.

[Fr] J. Frenkel, *Cohomologie non-abélienne et espaces fibrés*, Bull. Soc. Math. Fr. **85** (1957), 135-218.

[Gi] J. Giraud, *Cohomologie Non-Abélienne*, Grundl. Math. Wiss., vol. 179, Springer Verlag, Berlin-Heidelberg-New York, 1971.

[K-V] M. Kapranov and V. Voevodsky, *2-categories and Zamolodchikov's tetrahedra equations*, Proc. Symp. Pure Math., Part 2, vol. 56, 1994, pp. 177–259.

[P-S] A. Pressley and G. Segal, *Loop Groups*, Clarendon Press, Oxford, 1986.

[SGA1] A. Grothendieck et al., *Revêtements Etales et Groupe Fondamental, Séminaire de Géométrie Algébrique 1*, Lecture Notes in Math., vol. 224, Springer Verlag, Berlin-Heidelberg-New York, 1971.

DEPARTMENT OF MATHEMATICS, PENN STATE UNIVERSITY, UNIVERSITY PARK, PA 16802

E-mail address: jlb@math.psu.edu

Contemporary Mathematics
Volume **230**, 1998

The role of Michael Batanin's monoidal globular categories

ROSS STREET

ABSTRACT. This is a report on recent work of Michael Batanin [3] which replaces and extends [2]. The goal of his work is to provide an environment for defining the concepts associated with weak ω-categories and for developing the ensuing theory. The approach is "globular".

Lecture I: Globular categories and trees

BRIEF HISTORY

Some important steps in the development of weak ω-categories were as follows. Categories were defined by Eilenberg–Mac Lane in 1945. Monoidal and symmetric monoidal categories were defined by Mac Lane in 1963 (the finitely axiomatized concept is Mac Lane's; the name is that of Eilenberg-Kelly 1965). Ehresmann defined (strict) n-categories in 1966. Bénabou defined bicategories in 1967. In the early 80s, monoidal bicategories were in the air but a full definition was not published in that period. Joyal-Street defined braided monoidal categories in 1985. Gordon-Power-Street defined tricategories in 1991 (this, and the coherence theorem, were published in 1995). Braided monoidal bicategories were defined by Kapranov-Voevodsky-Baez-Neuchl-Breen around 1993. Trimble produced a definition of tetracategory in 1995.

Diverse approaches to weak n-categories for all n have appeared. Street (1985) suggested a simplicial definition with horn filler conditions. Trimble (1994) approached the problem using operads and Stasheff associahedra. Baez-Dolan (1995) have a definition using typed operads and opetopes (this is the topic of Baez's talk at this conference). Tamsamani (1996) gave a multisimplicial definition [6]. Batanin uses higher operads and globular sets.

FLAVOURS OF OPERAD

The table appearing below gives the steps in the development of three flavours of operad: Stasheff (or non-permutative) operads, the original May operads [4], and the higher operads of Batanin [2], [3]. In some sense, the whole theory in a given column develops once the top entry MC, SMC or MGC is declared. My two lectures will describe the third column in more detail. Batanin's construct of an operad whose algebras are weak ω-categories will also be explained.

1991 *Mathematics Subject Classification.* Primary 18D05; Secondary 55P35, 18D10, 55Q05.

This work was supported in part by an Australian Research Grant.

Stasheff	May	Batanin
monoidal category (MC) $\mathcal{V}$	symmetric monoidal category (SMC) $\mathcal{V}$	monoidal globular category (MGC) $\mathcal{V}$
free MC **N** on the terminal category $\mathcal{U}$	free symmetric monoidal category $\mathfrak{S}$ on the terminal category $\mathcal{U}$	free MGC **Tree** on the terminal GC $\mathcal{U}$
free MC $\mathbf{N}\int\mathcal{C}$ on any category $\mathcal{C}$	free SMC $\mathfrak{S}\int\mathcal{C}$ on any category $\mathcal{C}$	free MGC $D_s\mathcal{C}$ on any GC $\mathcal{C}$
functors $\mathbf{N} \longrightarrow \mathcal{V}$ "sequences"	functors $\mathfrak{S} \longrightarrow \mathcal{V}$ "Joyal species"	globular functors $\mathbf{Tree} \longrightarrow \mathcal{V}$ "collections"
substitution product of sequences (power series)	substitution product of species (Taylor series)	substitution tensor product of collections
an *operad* M is a monoid for the substitution tensor product		
$X : \mathcal{U} \longrightarrow \mathcal{V}$ X is an object of $\mathcal{V}$ $X^{\otimes n}$, $n \in \mathbf{N}$	$X : \mathcal{U} \longrightarrow \mathcal{V}$ X is an object of $\mathcal{V}$ $X^{\otimes n}$, $n \in \mathfrak{S}$	$X : \mathcal{U} \longrightarrow \mathcal{V}$ X is a globular object of $\mathcal{V}$ X^T, $T \in \mathbf{Tree}$
operad M gives a monad on good $\mathcal{V}$ $X \mapsto \sum_n M_n \otimes X^{\otimes n}$	operad M gives a monad on good $\mathcal{V}$ $X \mapsto \sum_n (M_n \otimes X^{\otimes n})/\mathfrak{S}_n$	operad M gives a monad on globular objects in good $\mathcal{V}$ $X \mapsto \sum_T A(T) \otimes X^T / \sim$
an M-*algebra* is an Eilenberg-Moore algebra for the monad		

WHAT ARE MONOIDAL GLOBULAR CATEGORIES ?

Let **Glob** denote the subcategory of **Set** consisting of the sets $\{0, n\}$, $n \geq 0$, and, apart from identity functions, only the constant functions $\{0, n\} \longrightarrow \{0, m\}$ with $n > m$. A *globular object* X of a category $\mathcal{X}$ is a functor $X : \mathbf{Glob} \longrightarrow \mathcal{X}$. We write X_n for $X\{0, n\}$ and $s_m, t_m : X_n \longrightarrow X_m$ for the image under X of the functions $\{0, n\} \longrightarrow \{0, m\}$ constant at 0, m, respectively.

$$\cdots \underset{t_3}{\overset{s_3}{\rightrightarrows}} X_3 \underset{t_2}{\overset{s_2}{\rightrightarrows}} X_2 \underset{t_1}{\overset{s_1}{\rightrightarrows}} X_1 \underset{t_0}{\overset{s_0}{\rightrightarrows}} X_0$$

The globular object X is called *n-skeletal* when $X_m = X_n$ and s_m, t_m are identities for all $m > n$. The globular object X is called *reflexive* when it is equipped with arrows $i : X_n \longrightarrow X_{n+1}$ for all $n \geq 0$ such that $s_n \circ i = t_n \circ i = 1_{X_n}$. We also write i or $i^k : X_n \longrightarrow X_{n+k}$ for the arrow obtained by composing k consecutive i arrows.

Suppose the category $\mathcal{X}$ has pullbacks. For $m < n$, define the object $X_n \underset{m}{\times} X_n$ to be the following pullback.

$$\begin{array}{ccc} X_n \underset{m}{\times} X_n & \xrightarrow{p_2} & X_n \\ {\scriptstyle p_1}\downarrow & & \downarrow{\scriptstyle t_m} \\ X_n & \xrightarrow[s_m]{} & X_m \end{array}$$

Recall that an *ω-category in* $\mathcal{X}$ is a reflexive globular object X together with arrows

$$\#_m : X_n \underset{m}{\times} X_n \longrightarrow X_n \qquad \text{for all } m < n$$

such that, for all $k < m < n$, the following induced 4×4 diagram is a truncated double simplicial object where each $X^{r,s}_{k,m,n}$ is the pullback of the previous outside pair of arrows.

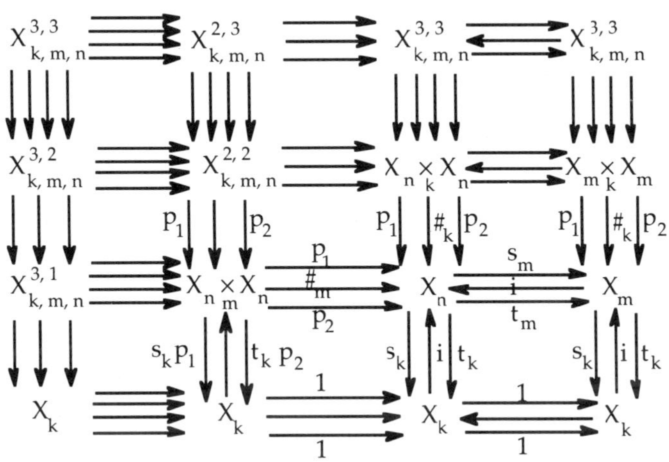

When $\mathcal{X}$ = **Set**, a globular object is called a *globular set* (or "ω-graph") and an ω-category in $\mathcal{X}$ is called an *ω-category*. We are also interested in the case $\mathcal{X}$ = **Cat** when a globular object is called a *globular category;* these form a 2-category **GC**. However, we are interested in something weaker than ω-categories in **Cat** – Batanin calls them *monoidal globular categories.* These are reflexive globular categories $\mathcal{C}$ together with functors $\#_m : \mathcal{C}_n \underset{m}{\times} \mathcal{C}_n \longrightarrow \mathcal{C}_n$ satisfying the equations

$$s_m(A \#_m B) = s_m A, \quad t_m(A \#_m B) = t_m A, \quad s_m(A \#_k B) = s_m A \#_k s_m B, \quad t_m(A \#_k B) = t_m A \#_k t_m B,$$

and globular natural isomorphisms

$$\alpha_m : (A \#_m B) \#_m C \xrightarrow{\ \cong\ } A \#_m (B \#_m C) \quad \textit{(associativity constraint)}$$

$$\lambda_m : i\, s_m A \#_m A \xrightarrow{\cong} A \quad \textit{(left unit constraint)}$$

$$\rho_m : A \#_m i\, s_m A \xrightarrow{\cong} A \quad \textit{(right unit constraint)}$$

$$\eta_{k,m} : (A \#_k B) \#_m (C \#_k D) \xrightarrow{\cong} (A \#_m C) \#_k (B \#_m D) \quad \textit{(interchange constraint)},$$

for all $k < m < n$ and all $A, B, C, D \in \mathcal{C}_n$ for which the expressions make sense, such that the obvious induced $4 \times 4 \times 4$ pseudo-diagram is a truncated triple pseudo-simplicial category. There is a 2-category **MGC** of monoidal globular categories; the arrows are globular functors $F : \mathcal{C} \longrightarrow \mathcal{D}$ equipped with natural isomorphisms $\phi_2 : FA \#_m FB \longrightarrow F(A \#_m B)$, $\phi_0 : iFA \longrightarrow FiA$ satisfying appropriate coherence conditions.

SOME EXAMPLES

1. An ω-category is precisely a monoidal globular category $\mathcal{C}$ for which each $\mathcal{C}_n$ is a discrete category.
2. An ω-category in **Cat** is precisely a strict monoidal globular category (that is, all the constraints are identities).
3. A bicategory is precisely a 1-skeletal monoidal globular category $\mathcal{C}$ for which $\mathcal{C}_0$ is a discrete category.
4. A double category is a 1-skeletal strict monoidal globular category.
5. A braided monoidal category is precisely a 2-skeletal monoidal globular category $\mathcal{C}$ for which $\mathcal{C}_0$ and $\mathcal{C}_1$ are equal to the terminal category **1**.
6. For each $n > 2$, a symmetric monoidal category is precisely an n-skeletal monoidal globular category $\mathcal{C}$ for which $\mathcal{C}_m = \mathbf{1}$ for all $m < n$.
7. Given a globular category $\mathcal{C}$, we can define its *suspension* to be the globular category $\mathcal{C}^+$ with $\mathcal{C}^+_0 = \mathbf{1}$, $\mathcal{C}^+_n = \mathcal{C}_{n-1}$ for $n > 0$ with s_m, t_m translated by 1 for $m > 0$, and with s_0, t_0 the only functors possible. A monoidal structure on $\mathcal{C}^+$ gives a monoidal structure on $\mathcal{C}$ by translating by 1; but it also gives extra functors $\otimes = \#_{-1} : \mathcal{C}_n \times \mathcal{C}_n \longrightarrow \mathcal{C}_n$ and objects $I_n \in \mathcal{C}_n$. We call the monoidal globular set $\mathcal{C}$ *augmented* when it is equipped with this extra structure making $\mathcal{C}^+$ monoidal. There is an obvious 2-category **AMGC** of augmented monoidal globular categories.

THE ALGEBRA OF TREES

The general operation, called *pasting,* available in an ω-category has received considerable attention. In order to discuss general pasting I introduced a very general concept of higher dimensional graph, called *computad.* The definition of n-computad is recursive: *a priori* the construction of the free (n–1)-category on an (n–1)-computad is required. It seems clear that some form of higher pasting should be involved in the definition of weak n-category. For example, as the main axiom for a tricategory looks like the fifth oriental or associahedron, it seems likely that we will need computads more general than ω-graphs. A useful kind of computad which can be defined directly, without recursion, is *parity complex* (or "pasting scheme"). These are a kind of generalised globular object in the category whose objects are finite sets and whose arrows are relations: but there are axioms which imply a lack of circuits ("loop freeness").

A major insight of Batanin was that a definition of weak ω-category could exist using the pasting generated by ω-graphs (= globular sets). This seems very restrictive since, in a 2-category, this would mean pasting only diagrams of 2-cells whose sources and targets were single arrows, not paths. As in many similar situations, this kind of pasting is controlled by how it behaves on the *terminal globular set.* That is, we must first understand the free ω-category **Tree** on the terminal globular set **1**: the globular set with exactly one element U_n in each dimension. This is as non loop free as one could get: *all* cells are endomorphisms. So we cannot resort to the construction of the free ω-category on a parity complex.

Clearly **Tree** has one 0-cell $U_0 = \bullet$ since there are no operations creating new 0-cells from old. The underlying 1-category of **Tree** is the free category on the terminal graph: it has one object U_0 and one arrow $U_1 \#_0 U_1 \#_0 \ldots \#_0 U_1 : U_0 \longrightarrow U_0$ for each natural number n; we denote this by what Trimble calls the *n-sprout:*

Then we have the 2-cell $U_2 : U_1 \Rightarrow U_1 : \bullet \longrightarrow \bullet$ which we denote by the tree

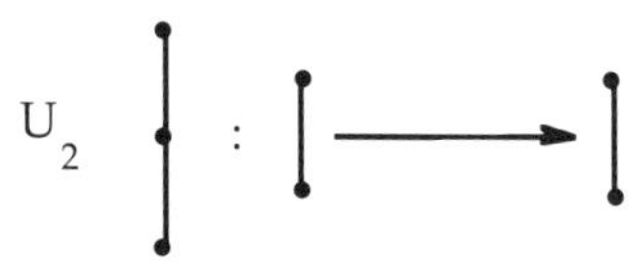

and can then whisker on the right by an n-sprout and on the left by an m-sprout: the result is an (m+1+n)-sprout with an extra edge going up from the (m+1)-st top vertex of the sprout.

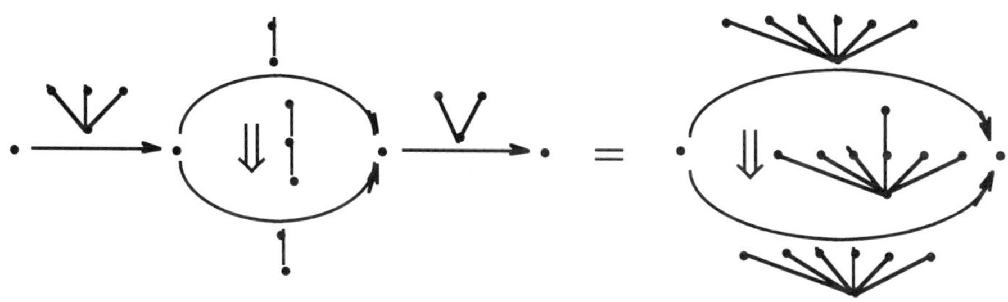

The composition $\#_1$ works as follows.

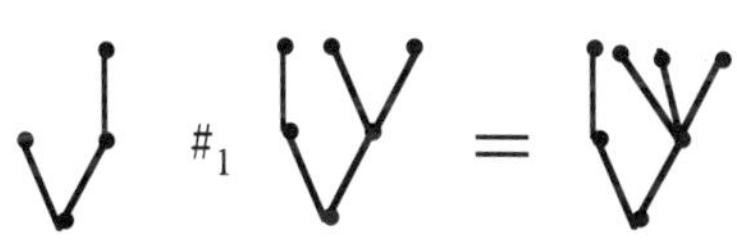

This leads us to the following model of the ω-category **Tree**.

We write *Ord* for the category whose objects are finite linearly ordered sets and whose arrows are order-preserving functions. Arrows $f : P \longrightarrow K$, $g : Q \longrightarrow K$ of *Ord* determine a linear order on the disjoint union $P+Q$ such that the inclusions $P \subseteq$

$P+Q$, $Q \subseteq P+Q$ are order preserving and, for $p \in P$, $q \in Q$, we have $p < q$ if and only if $f(p) \le g(q)$. We write $f \underset{K}{\oplus} g : P + Q \longrightarrow K$ for the function whose restriction to P, Q is f, g, respectively. This defines a monoidal structure $\underset{K}{\oplus}$ on each slice category $\mathcal{Ord}/K$. We write Δ for the full subcategory of $\mathcal{Ord}$ consisting of the ordinals $[n] = \{0, 1, \ldots, n\}$, including the empty ordinal $[-1]$.

An *n-stage tree* is a functor $T : [n]^{op} \longrightarrow \Delta$ such that $T(0) = [0]$. Elements of $T(i)$ are called *vertices of* T *of height* i. We obtain a planar graph from T by defining there to be an edge e from $v' \in T(i+1)$ to $v \in T(i)$ when v is the value of the function $T(i < i+1) : T(i+1) \longrightarrow T(i)$ at v'. Such an edge e is called an *input edge at* v. Vertices with no input edges are called *input vertices of* T or *leaves of* T. Write $\mathcal{Tr}_n$ for the functor category $[[n]^{op}, \mathcal{Ord}]$ whose isomorphism classes are n-stage trees[1]. Restriction along the inclusion $[m] \longrightarrow [n]$ for $m < n$ gives a functor $\partial_m : \mathcal{Tr}_n \longrightarrow \mathcal{Tr}_m$; and we make a globular category $\mathcal{Tr}$ by putting $s_m = t_m = \partial_m$. We make $\mathcal{Tr}$ reflexive by defining $i : \mathcal{Tr}_n \longrightarrow \mathcal{Tr}_{n+1}$ as follows:

$$(iT)(m) = T(m) \text{ for } m \le n, \qquad (iT)(n+1) = [-1].$$

Indeed, $\mathcal{Tr}$ becomes a monoidal globular category; the functors $\#_m : \mathcal{Tr}_n \underset{m}{\times} \mathcal{Tr}_n \longrightarrow \mathcal{Tr}_n$ are defined as follows[2]. For $S, T \in \mathcal{Tr}_n$ with $\partial_m S = \partial_m T$, define $R = S \#_m T$ by requiring $\partial_m R = \partial_m S$ and, for $m \le j \le k \le n$, putting

$$R(j \le k) = (\, S(j \le k) \underset{S(m)}{\oplus} T(j \le k) \,).$$

The ω-category **Tree** is defined by taking $\mathbf{Tree}_n$ to be the set of n-stage trees with the ω-category structure induced from the monoidal globular structure on $\mathcal{Tr}$.

We write U_n for the n-stage tree with one vertex of each height $m \le n$. The fundamental role played by **Tree** is explained by the next result which we have indicated above.

Theorem *The ω-category* **Tree** *is freely generated by its cells* U_n, $n \ge 0$.

FREE MONOIDAL GLOBULAR CATEGORIES

It is fairly clear that the forgetful functor ω-Cat(**Cat**) $\longrightarrow$ Glob(**Cat**) from strict monoidal globular categories to globular categories is monadic. In particular, it has a left adjoint

$$D_s : \text{Glob}(\mathbf{Cat}) \longrightarrow \omega\text{-Cat}(\mathbf{Cat}).$$

The explicit description of $D_s(\mathcal{C})$ is a crucial aspect of this work. If $\mathcal{C}$ is terminal in Glob(**Cat**), the Theorem of the last Section tells us that $D_s(\mathcal{C}) = \mathbf{Tree}$. So we expect trees to be involved in the description of $D_s(\mathcal{C})$ for a general globular category $\mathcal{C}$. Just as the free strict monoidal category on a category $\mathcal{X}$ has as objects the families of

[1] Notice that the only automorphisms in this category are identities.

[2] We are indebted to Marco Grandis for pointing out that our earlier description did not say what was intended.

objects of $\mathcal{X}$ indexed by the free strict monoidal category **N** on the terminal category, the objects of $D_s(\mathcal{C})$ are "families of objects of $\mathcal{C}$ indexed by trees" or "$\mathcal{C}$-labelled trees".

A *$\mathcal{C}$-labelling for a tree* $T \in \mathcal{T}r_n$ is a globular functor $\rho : T^* \longrightarrow \mathcal{C}$ where T^* is a globular set determined by T as follows. For vertices u, v of T, write $u \vee v$ for the highest vertex for which there exist directed paths $p : u \longrightarrow u \vee v$, $q : v \longrightarrow u \vee v$ of edges in T. The globular set T^* is generated by the leaves of T in dimensions equal to their height, subject only to the relations

$$t_m(u) = s_m(v)$$

for leaves u, v having $u \vee v \in T(m)$ and either

u, v are consecutive vertices of the same height, or

the paths p, q go through only maximal, minimal (respectively) edges of T at every height. It is in fact possible to reconstruct the tree T from the globular set T^*. Some examples of T^* are displayed below.

$U_2 =$  $U_2^* =$ 

$U_2 \#_0 U_2 =$  $(U_2 \#_0 U_2)^* =$ 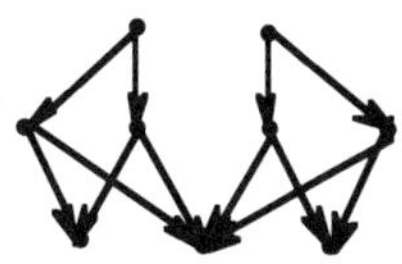

$U_2 \#_1 U_2 =$  $(U_2 \#_1 U_2)^* =$ 

Respectively, in these examples, globular functors $\rho : T^* \longrightarrow \mathcal{C}$ amount to globular pasting diagrams in $\mathcal{C}$ of the following shapes.

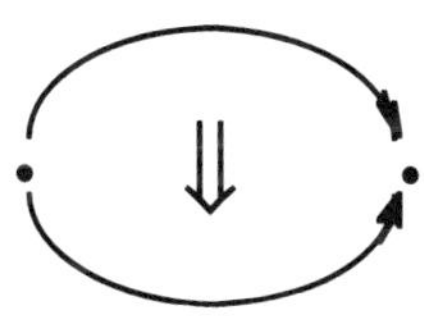

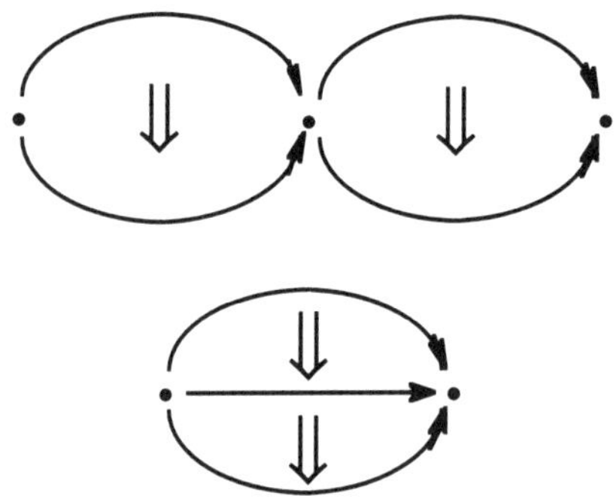

We define an *n-stage diagram in* $\mathcal{C}$ to be a pair (T, ρ) where T is an n-stage tree and σ is a $\mathcal{C}$-labelling for T. (Later we shall see how to assign a *value* $[\rho]$ to such a diagram when $\mathcal{C}$ is monoidal.)

The category $D_s(\mathcal{C})_n$ is the category whose objects are n-stage diagrams in $\mathcal{C}$ and whose arrows are pointwise globular natural transformations. The obvious inclusions

$$\sigma_m, \tau_m : (\partial_m U_n)^* = U_m^* \longrightarrow U_n^*$$

generate globular functions

$$\sigma_m, \tau_m : (\partial_m T)^* \longrightarrow T^*$$

for every n-stage tree T, and restriction along these gives the required globular category structure on $D_s(\mathcal{C})$. Since every n-stage diagram (T, ρ) can be regarded as an (n+1)-stage diagram (iT, ρ), we have our reflexive structure. For the monoidal structure we use the following pushouts of globular sets which exist for m-composable trees S, T.

$$\begin{array}{ccc} (i^m \partial_m S)^* & \xrightarrow{\sigma_m} & T^* \\ \downarrow{\scriptstyle \tau_m} & & \downarrow \\ S^* & \longrightarrow & (S \#_m T)^* \end{array}$$

Since diagrams (S, ρ), (T, ρ') are m-composable when $\rho\tau_m = \rho'\sigma_m$, the pushout yields a globular functor $\rho \#_m \rho' : (S \#_m T)^* \longrightarrow \mathcal{C}$; so we put $(S, \rho)\#_m (T, \rho') = (S\#_m T, \rho \#_m \rho')$. There is a globular functor $\upsilon : \mathcal{C} \longrightarrow D_s(\mathcal{C})$ which takes $C \in \mathcal{C}_n$ to the diagram (U_n, C) where C is identified with the globular functor $U_n^* \longrightarrow \mathcal{C}$ whose value at the unique highest node is C.

Theorem $D_s(\mathcal{C})$ *is the free strict monoidal globular category on a globular category* $\mathcal{C}$.

COHERENCE FOR MONOIDAL GLOBULAR CATEGORIES

By a familiar-style recursive construction, it is possible to describe the free monoidal globular category $D(\mathcal{C})$ on a globular category $\mathcal{C}$. The universal property of the inclusion $\mathcal{C} \longrightarrow D(\mathcal{C})$ of generators is that, for each monoidal globular category $\mathcal{D}$ and each globular functor $F : \mathcal{C} \longrightarrow \mathcal{D}$, there exists a unique strict monoidal globular functor $F' : D(\mathcal{C}) \longrightarrow \mathcal{D}$ whose restriction to $\mathcal{C}$ is F. In particular, there is a strict monoidal globular functor

$$\Gamma : D(\mathcal{C}) \longrightarrow D_s(\mathcal{C})$$

whose restriction to $\mathcal{C}$ is $\upsilon : \mathcal{C} \longrightarrow D_s(\mathcal{C})$.

Theorem *For all globular categories $\mathcal{C}$, the monoidal globular functor $\Gamma : D(\mathcal{C}) \longrightarrow D_s(\mathcal{C})$ is an equivalence. Every monoidal globular functor $D(\mathcal{C}) \longrightarrow \mathcal{D}$ is isomorphic to a strict monoidal globular functor $D(\mathcal{C}) \longrightarrow \mathcal{D}$.*

For any <u>monoidal</u> globular category $\mathcal{C}$, let $\mu : D(\mathcal{C}) \longrightarrow \mathcal{C}$ be the unique strict monoidal globular functor whose restriction to $\mathcal{C}$ is the identity. The composite of μ with an inverse equivalence for Γ gives a monoidal globular functor

$$[\ ] : D_s(\mathcal{C}) \longrightarrow \mathcal{C}$$

whose composite with $\upsilon : \mathcal{C} \longrightarrow D_s(\mathcal{C})$ is isomorphic to the identity of $\mathcal{C}$. This gives a *value* $[\rho] \in \mathcal{C}_n$ to each n-stage diagram (T, ρ) in $\mathcal{C}$. The following square commutes up to canonical isomorphism.

$$\begin{array}{ccc} D_s(D_s(\mathcal{C})) & \xrightarrow{[\ ]} & D_s(\mathcal{C}) \\ {\scriptstyle D_s([\ ])}\downarrow & & \downarrow{\scriptstyle [\ ]} \\ D_s(\mathcal{C}) & \xrightarrow[{[\ ]}]{} & \mathcal{C} \end{array}$$

Corollary *Every globular monoidal category $\mathcal{C}$ is equivalent to a strict one* **st**$(\mathcal{C})$ *which is the full image of $[\ ] : D_s(\mathcal{C}) \longrightarrow \mathcal{C}$.*

By applying these results to $\mathcal{C}^{+}$ we get coherence results for the augmented case.

<u>*Lecture II:* **Higher operads and weak ω-categories**</u>

THE ALGEBRA OF HIGHER SPANS

There is an augmented monoidal globular category **Span** in which all monoidal globular categories are "enriched". For each natural number n, let $\mathbf{Oct}_n$ be the poset

$$(\{0, 1\} \times \{0, 1, \ldots, n-1\}) \cup \{n\}$$

where n is the maximum element and $(\alpha, k) < (\beta, m)$ iff $k < m$. Define functions

$$\sigma_m , \tau_m \ : \ \mathbf{Oct}_m \longrightarrow \mathbf{Oct}_n$$

by $\sigma_m(\alpha, k) = \tau_m(\alpha, k) = (\alpha, k)$ for $k < m$, $\sigma_m(m) = (0, m)$, $\tau_m(m) = (1, m)$. An *n-span*[3] in a finitely complete category $\mathcal{X}$ is a functor $E : \mathbf{Oct}_n{}^{op} \longrightarrow \mathcal{X}$ and a morphism of n-spans is a natural transformation; let $\mathbf{Span}(\mathcal{X})_n$ denote this functor category.

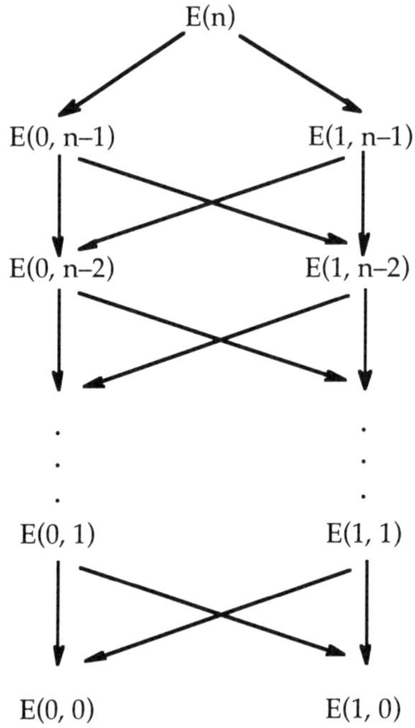

Restriction along σ_m , τ_m gives the globular category structure s_m , t_m on the sequence $\mathbf{Span}(\mathcal{X})$ of categories $\mathbf{Span}(\mathcal{X})_n$, $n \geq 0$. For each n-span E, the (n+1)-span iE is given by $iE(\alpha, k) = E(\alpha, k)$ for $k < n$ and $iE(\alpha, n) = iE(n+1) = E(n)$; this makes $\mathbf{Span}(\mathcal{X})$ reflexive. If E, F are m-composable n-spans then the composite $E \#_m F$ is given by the pullbacks

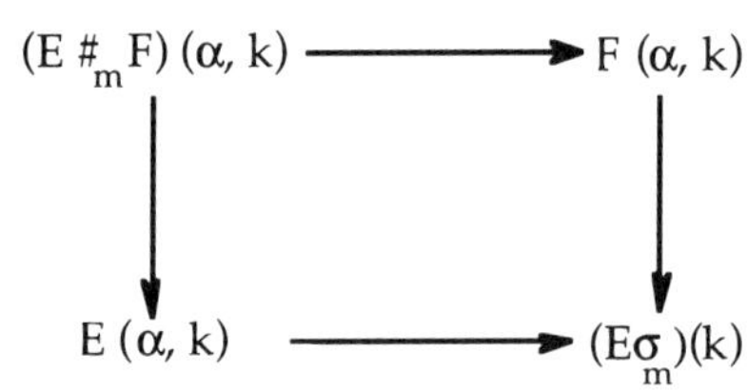

and with the m-source, m-target of $E \#_m F$ equal to $s_m E$, $t_m F$, respectively. With the

[3] In Paris, 1994, P. Cartier gave a lecture on 'multicategories' where n-spans were called 'telescopes of height n'. Stephen Lack kept notes.

canonical constraints coming from the universal property of pullback, **Span**($\mathcal{X}$) becomes a monoidal globular category. Taking $\#_{-1} = \otimes$ to be binary categorical product, we see that **Span**($\mathcal{X}$) becomes augmented. (In fact, this augmenting process can be iterated: **Span**$(\mathcal{X})^{+}$ becomes augmented by again taking $\#_{-1}$ to be product.)

Each globular object of any category $\mathcal{X}$ leads to an n-span in $\mathcal{X}$ for all $n \geq 0$. More precisely, there is a canonical functor $\mathbf{Oct}_n{}^{\mathrm{op}} \longrightarrow \mathbf{Glob}$ taking n to $\{0, n\}$ and (α, m) to $\{0, m\}$; restriction along this functor turns the globular object X in $\mathcal{X}$ into an n-span.

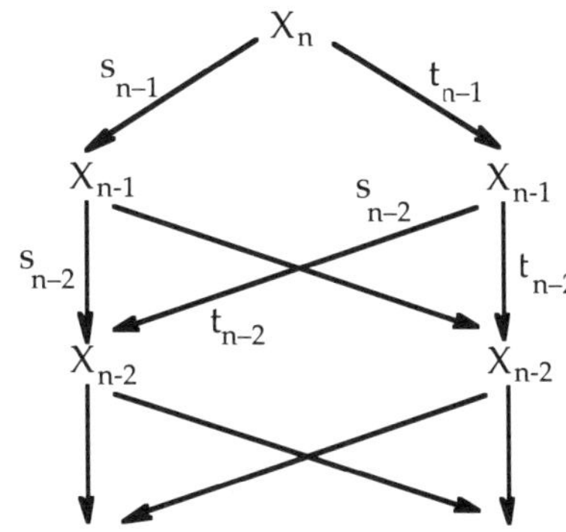

Merely write **Span** for the augmented monoidal globular category **Span(Set)**. Suppose $\mathcal{C}$ is any globular category. For all objects A, B of $\mathcal{C}_n$, we obtain the n-span $\mathrm{HOM}_{\mathcal{C}}(A, B)$ as follows.

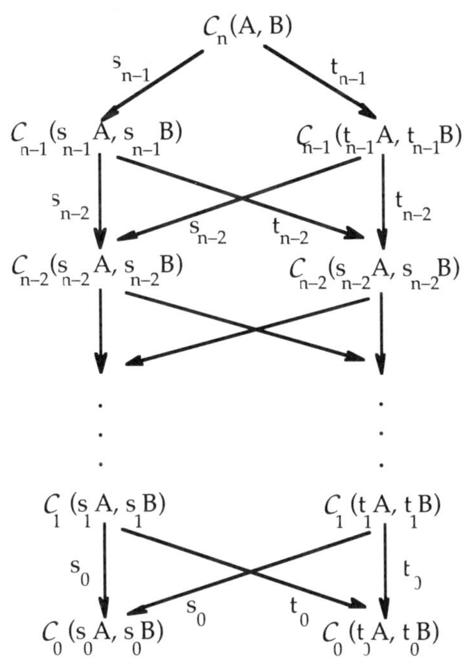

For a category $\mathcal{X}$ with finite colimits, there is the augmented globular category $\mathbf{Cospan}(\mathcal{X}) = \mathbf{Span}(\mathcal{X}^{op})^{op}$.

KAN EXTENSIONS OF GLOBULAR FUNCTORS

We shall review some notions from [5] in the case of categories varying over the category $\mathbf{Glob}^{op}$. This viewpoint of Batanin's work will be developed more fully in a future paper.

A category $\mathcal{C}$ varying over $\mathbf{Glob}^{op}$ is just a category-valued functor on $\mathbf{Glob}$; that is, a globular category. We call $\mathcal{C}$ *cocomplete* when each $\mathcal{C}_n$ is a cocomplete category, each of the functors $s_m, t_m : \mathcal{C}_n \longrightarrow \mathcal{C}_m$ has a left adjoint $s'_m, t'_m : \mathcal{C}_m \longrightarrow \mathcal{C}_n$, and the Chevalley-Beck condition is satisfied (this takes a little care in stating [5; §7.11] since $\mathbf{Glob}$ does not have pullbacks; the precise details do not concern us now). An example of a cocomplete globular category is $\mathbf{Span}$.

If $\mathcal{C}$ is a cocomplete globular category then, for all globular functors $J : \mathcal{A} \longrightarrow \mathcal{B}$ where $\mathcal{A}$ is "small" and $\mathcal{B}$ has "small homs", each globular functor $F : \mathcal{A} \longrightarrow \mathcal{C}$ has a *left Kan extension* $K = \mathrm{Lan}_J F : \mathcal{B} \longrightarrow \mathcal{C}$; we shall give the formula[4].

Take $B \in \mathcal{B}_n$. Let $n\#\mathcal{A}$ denote the result of applying the Grothendieck construction to the composite functor

$$\mathbf{Oct}_n^{op} \longrightarrow \mathbf{Glob} \xrightarrow{\mathcal{A}} \mathbf{Cat}\,.$$

All objects of the category $n\#\mathcal{A}$ are either of the form (n, A) with $A \in \mathcal{A}_n$ or of the form (α, m, A) with $0 \le m < n$, $\alpha = 0, 1$, $A \in \mathcal{A}_m$. Define a functor $\mathcal{B}(J, B) : (n\#\mathcal{A})^{op} \longrightarrow \mathbf{Set}$ by the assignments

$$(n, A) \longmapsto \mathcal{B}_n(J_n A, B), \qquad (0, m, A) \longmapsto \mathcal{B}_m(J_m A, s_m B),$$
$$(1, m, A) \longmapsto \mathcal{B}_m(J_m A, t_m B).$$

Define a functor $F^{/n} : n\#\mathcal{A} \longrightarrow \mathcal{C}_n$ by the assignments

$$(n, A) \longmapsto F_n A, \qquad (0, m, A) \longmapsto s'_m F_m A, \qquad (1, m, A) \longmapsto t'_m F_m A.$$

Then the value of $K = \mathrm{Lan}_J F : \mathcal{B} \longrightarrow \mathcal{C}$ at B is given by the coend

$$K_n B = \int^{P \in n\#\mathcal{A}} \mathcal{B}(J, B)P \bullet F^{/n} P$$

where, for a set E and an object $C \in \mathcal{C}_n$, the object $E \bullet C$ of $\mathcal{C}_n$ is the coproduct of E copies of C.

COLLECTIONS

A *collection* in a globular category $\mathcal{C}$ is a globular functor $A : \mathbf{Tree} \longrightarrow \mathcal{C}$. Write $\mathbf{Coll}(\mathcal{C}) = \mathbf{GC}(\mathbf{Tree}, \mathcal{C})$ for the category of collections in $\mathcal{C}$.

If $\mathcal{C}$ is merely $\mathcal{V}^+$ for a monoidal category $\mathcal{V}$ then a collection in $\mathcal{C}$ is just a sequence of objects A_n, $n \ge 0$, of $\mathcal{V}$. These are the correct objects to support the usual

[4] We are really only interested in the case where $\mathcal{A}$ is a globular set.

structure of (non-permutative) operad on $\mathcal{V}$.

To support Batanin's higher operads he requires collections in spans. At this point notice that the category of collections in **Span** is isomorphic to the category of **Set**-valued functors on the category whose objects are trees $T \in \mathbf{Tree}_n$ and whose arrows are generated by formal symbols $s, t : T \longrightarrow \partial T$ subject to the usual source-target relations for successive pairs.

For any cocomplete augmented monoidal globular category $\mathcal{C}$ (such as **Span**), we shall now define the *substitution tensor product* $A \circ B$ of collections $A, B : \mathbf{Tree} \longrightarrow \mathcal{C}$. We shall refer to the two globular functors

$$\delta : D_s(\mathbf{Tree}) \longrightarrow \mathbf{Tree}, \quad [\ \ ] : D_s(\mathbf{Tree}) \longrightarrow \mathbf{Tree}$$

where δ assigns the parametrizing tree $\delta(T, \rho) = T$, and $[\ \ ]$ assigns the value $[\rho]$, to each diagram (T, ρ) of trees. Let $[B] : D_s(\mathbf{Tree}) \longrightarrow \mathcal{C}$ denote the (unique up to isomorphism) monoidal globular functor whose restriction to **Tree** is B. We also have $A\delta : D_s(\mathbf{Tree}) \longrightarrow \mathcal{C}$. Form the external tensor product $A\delta \otimes [B] : D_s(\mathbf{Tree}) \longrightarrow \mathcal{C}$ given by the pointwise formula

$$(A\delta \otimes [B])(T, \rho) = AT \otimes [B](T, \rho).$$

Here $\otimes$ is the augmented tensor product $\#_{-1}$ of $\mathcal{C}$. Define $A \circ B : \mathbf{Tree} \longrightarrow \mathcal{C}$ to be the left Kan extension of the globular functor $A\delta \otimes [B] : D_s(\mathbf{Tree}) \longrightarrow \mathcal{C}$ along $[\ \ ] : D_s(\mathbf{Tree}) \longrightarrow \mathbf{Tree}$.

Theorem *Substitution tensor product of collections is a monoidal structure on* **Coll**$(\mathcal{C})$.

Example Consider the case where $\mathcal{C} = \mathcal{V}^+$ for a cocomplete closed braided monoidal category $\mathcal{V}$. Then globular functors $F : \mathcal{G} \longrightarrow \mathcal{C}$ are determined by their 1-dimensional components $F_1 : \mathcal{G}_1 \longrightarrow \mathcal{V}$. A 1-stage tree T can be identified with the number n of vertices of T of height 1. So we can identify collections $A : \mathbf{Tree} \longrightarrow \mathcal{V}^+$ with sequences $(A_n)_{n \geq 0}$ of objects of $\mathcal{V}$ via the specification

$$A_n = A(T).$$

The Kan extension of globular functors along $[\ \ ] : D_s(\mathbf{Tree}) \longrightarrow \mathbf{Tree}$ reduces here to the Kan extension of ordinary functors along the component $[\ \ ]_1$ defined on the 1-stage diagrams. The value of the Kan extension at $T \in \mathbf{Tree}$ is therefore given by taking the coproduct of the terms $A(S) \otimes [B](S,\tau)$ as (S,τ) runs over the 1-stage diagrams with $[\tau] = T$. A 1-stage diagram $\tau : S^* \longrightarrow \mathbf{Tree}$ of trees with $[\tau] = T$ can be identified with a partition

$$n = j_1 + \ldots + j_k\,.$$

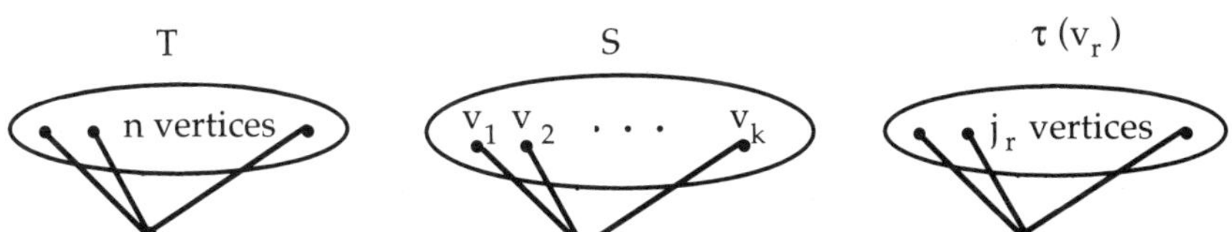

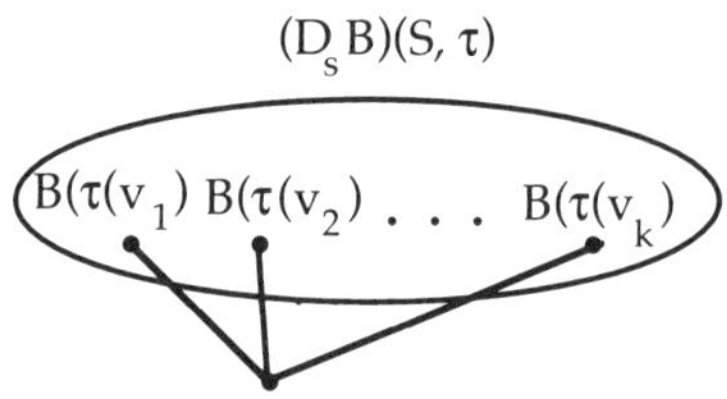

We can see that

$$[B](S, \tau) = B(\tau(v_1)) \otimes B(\tau(v_2)) \otimes \ldots \otimes B(\tau(v_k)) = B_{j_1} \otimes \ldots \otimes B_{j_k} \quad \text{and}$$

$$(A\,\delta)(S,\tau) = A(S) = A_k.$$

So we recapture the formula for the substitution tensor product needed for defining the usual (non-permutative) operads, namely,

$$(A \circ B)_n = \sum_{\substack{k \geq 0 \\ j_1 + \ldots + j_k = n}} A_k \otimes B_{j_1} \otimes \ldots \otimes B_{j_k}\,.$$

HIGHER OPERADS AND THEIR ALGEBRAS

Let $\mathcal{C}$ be a cocomplete augmented monoidal globular category.

Definition An *operad* in $\mathcal{C}$ is a monoid in the monoidal category **Coll**($\mathcal{C}$).

Write **Oper**($\mathcal{C}$) for the category of operads in $\mathcal{C}$. We are particularly interested in

Coll = **Coll(Span)** and **Oper** = **Oper(Span)**.

Theorem *With reasonable conditions on* $\mathcal{C}$ *(satisfied in the case* $\mathcal{C}$ = **Span**), *the forgetful functor* **Oper**($\mathcal{C}$) $\longrightarrow$ **Coll**($\mathcal{C}$) *is monadic and* **Oper**($\mathcal{C}$) *is cocomplete.*

The composite of **Oper**($\mathcal{C}$) $\longrightarrow$ **Coll**($\mathcal{C}$) with its left adjoint is the monad denoted by

$$\mathcal{F} : \mathbf{Coll}(\mathcal{C}) \longrightarrow \mathbf{Coll}(\mathcal{C})$$

and is given by $\mathcal{F}(A) = \sum_{n \geq 0} A^{\circ n}$. A *pointed* collection is a collection A together with a morphism $e : I \longrightarrow A$. Write $\mathbf{Coll}_p$ for the category of pointed collections. It is an easy consequence of the Theorem that the forgetful functor **Oper** $\longrightarrow$ $\mathbf{Coll}_p$ is monadic; the monad generated here is denoted by

$$\mathcal{F}_p : \mathbf{Coll}_p \longrightarrow \mathbf{Coll}_p.$$

A *globular object* X of a globular category $\mathcal{G}$ is a globular functor W : **1**

$\longrightarrow \mathcal{G}$ from the terminal globular category $\mathbf{1}$[5]. The globular objects in **Span** can be identified with globular sets. Write **Glob**($\mathcal{G}$) for the category of globular objects in $\mathcal{G}$.

By the freeness of **Tree**, each globular object $W : \mathbf{1} \longrightarrow \mathcal{C}$ in a monoidal globular category $\mathcal{C}$ extends to a unique up to isomorphism monoidal globular functor $[W] : \mathbf{Tree} \longrightarrow \mathcal{C}$. Globular objects in monoidal globular categories are exponentiated by trees $T \in \mathbf{Tree}$ according to the formula:

$$W^T = [W](T).$$

For example, if $\mathcal{C} = \mathcal{V}^+$ where $\mathcal{V}$ is a monoidal category, then a globular object W in $\mathcal{C}$ is just an object W of $\mathcal{V}$ and exponentiation gives the tensor power

$$W^T = W \otimes \ldots \otimes W \quad (\text{n terms})$$

where T is a 1-stage tree with n leaves.

Proposition *For every globular object* W *in* $\mathcal{C}$ *there is a canonical structure of operad in* **Span** *on the collection* $E_{\mathcal{C}}(W)$ *given by*

$$E_{\mathcal{C}}(W)(T) = \mathrm{HOM}_{\mathcal{C}}(W^T, W_n) \quad \textit{for } T \in \mathbf{Tree}_n.$$

Suppose A is an operad in **Span** and $\mathcal{C}$ is a monoidal globular category[6]. An A-*algebra in* $\mathcal{C}$ is a globular object W of $\mathcal{C}$ together with an operad morphism

$$\phi : A \longrightarrow E_{\mathcal{C}}(W).$$

An A-*algebra morphism* $f : V \longrightarrow W$ is a globular object morphism such that the following square commutes for all $T \in \mathbf{Tree}_n$.

$$\begin{array}{ccc} A(T) & \xrightarrow{\ \phi_T\ } & E_{\mathcal{C}}(V)(T) \\ {\scriptstyle \phi_T}\downarrow & & \downarrow{\scriptstyle \mathrm{HOM}_{\mathcal{C}}(V^T, f_n)(T)} \\ E_{\mathcal{C}}(W)(T) & \xrightarrow[\mathrm{HOM}_{\mathcal{C}}(f^T, W_n)(T)]{} & \mathrm{HOM}_{\mathcal{C}}(V^T, W_n)(T) \end{array}$$

In fact, each operad A in **Span** determines a monad on **Glob**($\mathcal{C}$) whose algebras are precisely the A-algebras. For defining ω-categories we only need this in the case where $\mathcal{C} = \mathbf{Span}$. To define the monad we need to regard globular sets as collections via the fully faithful functor

$$\Psi : \mathbf{GS} = \mathbf{GC}(\mathbf{1}, \mathbf{Span}) \longrightarrow \mathbf{GC}(\mathbf{Tree}, \mathbf{Span}) = \mathbf{Coll}$$

given by left Kan extension along the globular functor $\mathbf{1} \longrightarrow \mathbf{Tree}$ which, in dimension n, picks out the 0-stage tree U_0 regarded as an n-stage tree iU_0.

---

[5] So globular objects are global elements.

[6] For now we resist the temptation to define A-algebras in a "$\mathcal{C}$-category" where A is an operad in $\mathcal{C}$.

Proposition *Each operad* A *in* **Span** *determines a finitary monad* A *on the category* **GS** = [**Glob** , **Set**] *of globular sets via the formula*

$$\Psi(A(X)) = A \circ \Psi(X)$$

for all globular sets X. *The algebras for the operad* A *are precisely the Eilenberg-Moore algebras for the monad* A. *Thus, for example, the category of* A*-algebras is cocomplete.*

The main point is that, for each collection A : **Tree** ⟶ **Span**, the functor $A \circ -$ takes collections in the image of Ψ to collections in the image of Ψ.

One can modify these definitions in a fairly straightforward way to yield the notion and theory of *n-operads* which are monoids in the monoidal category of *n-collections.*

Examples 1) A 0-operad is a monoid in the monoidal category $\mathcal{C}_0$ with $\#_0$ as tensor product.

2) If $\mathcal{V}$ is a braided monoidal category then 1-operads in $\mathcal{V}^+ \in$ **AMGC** are precisely non-permutative operads in $\mathcal{V}$.

3) The terminal collection $\mathcal{M}$ is naturally an operad. The category of $\mathcal{M}$-algebras is isomorphic to ω-**Cat**.

4) There is an easily described 2-operad in **Span** whose algebras are derivation schemes.

5) There are two fairly easily described 2-operads in **Span** whose algebras are sesqui-categories and bicategories. There is an easy 3-operad in **Span** which gives Gray-categories as its algebras.

CONTRACTIBILITY

A globular set X is called *contractible in dimension* $n \geq 0$[7] when, for all $x, y \in X_n$ with

$$s_{n-1}(x) = s_{n-1}(y), \qquad t_{n-1}(x) = t_{n-1}(y),$$

there exists $z \in X_{n+1}$ with

$$s_n(z) = x, \qquad t_n(z) = y.$$

A choice function $(x,y) \longmapsto z$ is called a *contraction.*

A collection A : **Tree** ⟶ **Span** is called *contractible* when, for all $T \in \mathbf{Tree}_n$, the globular set

$$G_T(A) : \quad \cdots \rightrightarrows A(iT)(n+1) \rightrightarrows A(T)(n) \rightrightarrows \cdots \rightrightarrows A(\partial_0 T)(0)$$

is contractible in all dimensions $m \geq n$. An operad is *contractible* when its underlying collection is contractible.

An operad A is *normalized* when $A(U_0) = I_0$.

We need to distinguish a particular collection P : **Tree** ⟶ **Span**. Write 0^k_n for the n-span made up of all singleton sets 1 for $k = n$ and, for $k < n$, the description

[7] By convention, X_{-1} is taken to be a singleton.

of 0_n^k is

$$0_n^k(\alpha, m) = \begin{cases} 1 & \text{for } m \leq k \\ 0 & \text{for } k < m < n \end{cases}$$

$$0_n^k(n) = 0 .$$

Now define P by $P(T) = 0_n^{n-h}$ if there exists a $1 \leq k \leq h$ such that $\partial_h T = U_h \#_k U_h$ but there exists no $1 \leq k \leq h$ such that $\partial_{h+1} T = U_{h+1} \#_k U_{h+1}$. Let $\rho : I \longrightarrow P$ be the canonical morphism.

A *system of compositions* for an operad A is a collection morphism $m : P \longrightarrow A$ such that $m\,\rho = \varepsilon$. The idea is that such an operad A is equipped with the "compositions" required of a weak ω-category but not necessarily with any of the "constraints".

Lemma *There exists a pointed endofunctor* $C : \mathbf{Coll} \longrightarrow \mathbf{Coll}$, $\eta : 1 \longrightarrow C$, *such that* $C(X)$ *is contractible for all* $X \in \mathbf{Coll}$, *and, for every* $f : X \longrightarrow Y$ *in* **Coll** *with* Y *contractible, there exists* $f_c : C(X) \longrightarrow Y$ *with* $f = f_c\,\eta$.

Theorem *The category of contractible operads equipped with a system of compositions has a weak initial object* K.

Construction Put $A_0 = P$ and $K_0 = \mathcal{F}_p(A_0)$ (with identity morphism called ϕ_0). We have the canonical inclusion $i_0 : A_0 \longrightarrow K_0$. We certainly now have an operad K_0 with a system of compositions, but it is not contractible. Enter the Lemma: we have $\eta_0 : K_0 \longrightarrow C(K_0) = A_1$. Let $i_1 : A_1 \longrightarrow \mathcal{F}_p(A_1)$ be the canonical inclusion. Put $d_0 = i_1\,\eta_0$ and $d_1 = \mathcal{F}_p(j_0) : \mathcal{F}_p(A_0) \longrightarrow \mathcal{F}_p(A_1)$. Notice that d_1 is an operad morphism while d_0 is not. Define $\mathcal{F}_p(A_1) \longrightarrow K_1$ to be the coequalizer (in the category of operads) of the operad-morphism extensions $\mathcal{F}_p(\mathcal{F}_p(A_0)) \longrightarrow \mathcal{F}_p(A_1)$ of d_0, d_1. Clearly we have an operad morphism $K_0 \longrightarrow K_1$. This process iterates. (See the diagram on the next page.) Put $K = \operatorname{colim} K_n$ in the category of operads; it is clearly contractible with a system of compositions. An easy argument shows it is weakly initial among such operads.

PUNCH LINE

A *weak ω-category* is a K-algebra in **Span**.

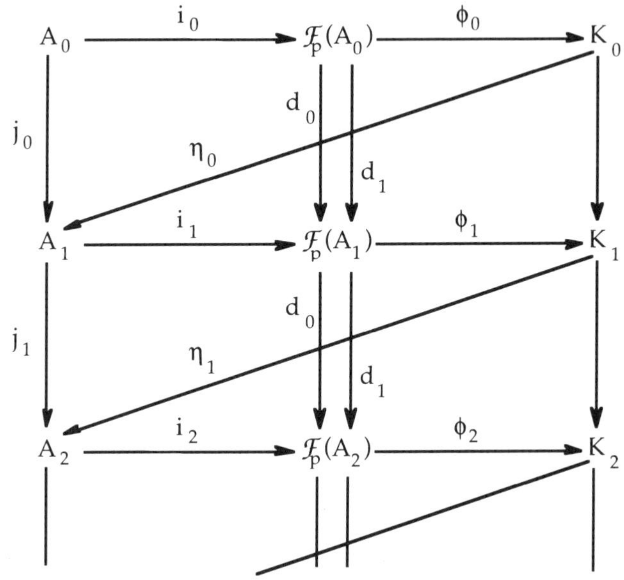

REFERENCES

[1] M.A. Batanin, Homotopy coherent homotopy theory and A_∞-structures in monoidal categories, *Journal of Pure & Applied Algebra* (to appear).

[2] M.A. Batanin, On the definition of weak ω-category, *Macquarie Mathematics Report* 96/207 (1996) 24pp.

[3] M.A. Batanin, Monoidal globular categories as a natural environment for the theory of weak n-categories, *Advances in Math.* 136 (1998) 39-103.

[4] P. May, *The Geometry of Iterated Loop Spaces,* Lecture Notes in Math. 271 (Springer-Verlag, 1972).

[5] R. Street, Conspectus of variable categories, *J. Pure Appl. Algebra* 21 (1981) 307-338.

[6] Z. Tamsamani, *Sur des notions de n-categorie et n-groupoide non-stricte via des ensembles multi-simpliciaux* (Thesis, Université Paul Sabatier, Toulouse, 1996: available on alg-geom 95-12 and 96-07).

Macquarie University
N S W 2109
AUSTRALIA
Email <street@mpce.mq.edu.au>

Contemporary Mathematics
Volume **230**, 1998

Braided deformations of monoidal categories and Vassiliev invariants

David N. Yetter

ABSTRACT. It is the purpose of this paper to provide an exposition of a cohomological deformation theory for braided monoidal categories, an exposition and proof of a very general theorem of the form "all quantum invariants are Vassiliev invariants," and to relate to the existence of the Vassiliev invariants obtained via the theorem to the vanishing of obstructions in the deformation theory.

1. Introduction

Our deformation theory for braided monoidal categories will be based on a deformation theory for monoidal functors. Remarkably, the deformation theory of monoidal functors, unlike the deformation theory of monoidal categories developed in [**3**] shares many of the properties of the Gerstenhaber[**8, 9, 10**] deformation theory of algebras, though the relationship between the two theories is unclear.[1] In this present paper, the deformation theory of monoidal functors is worked out only in the case where the monoidal structure of the source and target categories are undeformed.

Previous theorems of the "all quantum invariants are Vassiliev" type have involved definitions of "quantum invariants" either in terms of the well-known knot polynomial of Jones [**14**], HOMFLY [**12**], or Kauffman [**16**], as in the case of Birman and Lin [**2**] and Stanford [**21**], or more generally in terms of constructions tied explicitly to a (simple) Lie algebra, as in the case of Piunikhin [**19**]. Still, the notion of "quantum invariants" used here is not the most general conceivable, so some preliminaries are in order. The most general reasonable notion of "quantum invariant" of (framed) knots and links is what will here be called a "functorial invariant", that is, an R-valued invariant (R some commutative ring) which arises by choosing an object in an R-linear tortile (a.k.a. ribbon) category (with $End(\mathbf{I}) = R$, where $\mathbf{I}$ is the monoidal identity object), and considering the image of all framed links under

1991 *Mathematics Subject Classification.* Primary 18D10; Secondary 57M25, 14D15.
Supported by NSF Grant # DMS-9504423.

[1]It has been recently brought to the author's attention that some aspects of this work have been discovered independently by Davydov [**3a**]. Davydov's work does not however deal with "higher-order infinitesimal deformations", and thus the full parallel with Gerstenhaber's deformation theory is unrealized.

the functor induced by the coherence theorem of Shum [20]. If we wish to deal with unframed links in this context, we must find a normalization procedure. (One will always be available if the chosen object is simple.)

It is not the case that all functorial invariants are Vassiliev invariants: the "counting invariants" associated to finite groups are functorial (cf. Freyd and Yetter [6] and Yetter [24]), but as shown by Altschuler [1] are not generally Vassiliev. Indeed, even the functorial invariants usually shown to "be" Vassiliev invariants are not themselves Vassiliev invariants, but rather become a sequence of Vassiliev invariants when a suitable change of variable by substitution of a power series is applied. Our theorem will be stated in terms of those functorial invariants which arise from tortile (ribbon) deformations of a rigid symmetric monoidal category.

Our setting is more general than that used for previous results of this type, and, we have reason to conjecture, may be the most general setting in which a theorem of this type can be proved. We will point to examples of knot invariants already existing in the literature to show the greater generality. More importantly, by proving the result in a purely categorical setting, we are able to relate the Vassiliev invariants thus obtained to a new cohomology theory which both classifies and provides the obstruction theory for deformations of braidings on tensor categories. Once the relationship between our deformation theory and the Gerstenhaber deformation theory of algebras is clearer, it is reasonable to think that we will gain an improved understanding of the relation between quantum groups and Vassiliev theory.

We begin with a discussion of the categorical machinery necessary for the statement and proof of the theorem.

2. Tortile Categories and Deformations of Tensor Categories and Functors

The setting in which our theorem is proved involves a class of categories with structure which includes both the categories of representations of quantum groups and various categories of tangles. To make this explicit:

DEFINITION 2.1. *A* monoidal category $\mathcal{C}$ *is a category* $\mathcal{C}$ *equipped with a functor* $\otimes : \mathcal{C}\times\mathcal{C} \to \mathcal{C}$ *and an object* I*, together with natural isomorphisms* $\alpha : \otimes(\otimes \times 1_{\mathcal{C}}) \Rightarrow \otimes(1_{\mathcal{C}} \times \otimes)$, $\rho : \otimes I \Rightarrow 1_{\mathcal{C}}$, *and* $\lambda : I\otimes \Rightarrow 1_{\mathcal{C}}$, *satisfying the pentagon, triangle, and bigon* ($\rho_I = \lambda_I$) *coherence conditions (cf.* [17]*). A* tensor category over K, *for* K *some field (or commutative ring) is a monoidal abelian category linear over* K, *with all* $- \otimes X$ *and* $X \otimes -$ *exact. Similarly, a* semigroupal category *is a category equipped with only* $\otimes$ *and* α *satisfying the usual pentagon.*

DEFINITION 2.2. *A* (strong) monoidal functor $F : \mathcal{C} \to \mathcal{D}$ *between two monoidal categories* $\mathcal{C}$ *and* $\mathcal{D}$ *is a functor* F *between the underlying categories, equipped with a natural isomorphism*

$$\tilde{F} : F(- \otimes -) \to F(-) \otimes F(-)$$

and an isomorphism $F_0 : F(I) \to I$, *satisfying the hexagon and two squares*

$$
\begin{array}{ccc}
F((A \otimes B) \otimes C) & \xrightarrow{F(\alpha)} & F(A \otimes (B \otimes C)) \\
\tilde{F}_{A\otimes B,C}\big\downarrow & & \big\downarrow \tilde{F}_{A,B\otimes C} \\
F(A \otimes B) \otimes F(C) & & F(A) \otimes F(B \otimes C) \\
\tilde{F}_{A\otimes B} \otimes F(C)\big\downarrow & & \big\downarrow F(A)\otimes \tilde{F}_{B\otimes C} \\
(F(A) \otimes F(B)) \otimes F(C) & \xrightarrow[\alpha]{} & F(A) \otimes (F(B) \otimes F(C))
\end{array}
$$

$$
\begin{array}{ccc}
F(I \otimes A) & \xrightarrow{F(\lambda)} & F(A) \\
\tilde{F}_{I,A}\big\downarrow & & \big\uparrow \lambda \\
F(I) \otimes F(A) & \xrightarrow[F_0 \otimes F(A)]{} & I \otimes F(A)
\end{array}
\qquad
\begin{array}{ccc}
F(A \otimes I) & \xrightarrow{F(\rho)} & F(A) \\
\tilde{F}_{A,I}\big\downarrow & & \big\uparrow \rho \\
F(A) \otimes F(I) & \xrightarrow[F(A) \otimes F_0]{} & F(A) \otimes I
\end{array}
$$

A pair of a functor F and a natural isomorphism $\tilde{F}$ (without the isomorphism F_0 as above will be called a semigroupal functor.

DEFINITION 2.3. *A* monoidal natural transformation *is a natural transformation $\phi : F \Rightarrow G$ between monoidal functors which satisfies*

$$\tilde{G}_{A,B}(\phi_{A\otimes B}) = \phi_A \otimes \phi_B(\tilde{F}_{A,B})$$

and $F_0 = G_0(\phi_I)$. A semigroupal natural transformation *between semigroupal functors is defined similarly.*

We will be concerned with monoidal categories with additional structure.

DEFINITION 2.4. *A* braided monoidal category *is a monoidal category equipped with a monoidal natural isomorphism $\sigma : \otimes \Rightarrow \otimes(tw)$, where $tw : \mathcal{C} \times \mathcal{C} \to \mathcal{C} \times \mathcal{C}$ is the "twist functor" ($tw(f, g) = (g, f)$), satisfying*

$$
\begin{array}{ccccc}
 & A \otimes (B \otimes C) & \xrightarrow{\sigma^{\pm 1}} & (B \otimes C) \otimes A & \\
{}^{\alpha}\nearrow & & & & \searrow^{\alpha} \\
(A \otimes B) \otimes C & & & & B \otimes (C \otimes A) \\
{}_{\sigma^{\pm 1} \otimes C}\searrow & & & & \nearrow_{B \otimes \sigma^{\pm 1}} \\
 & (B \otimes A) \otimes C & \xrightarrow[\alpha]{} & B \otimes (A \otimes C) &
\end{array}
$$

A braided monoidal category is a symmetric monoidal category *if the components of σ satisfy $\sigma_{B,A}(\sigma_{A,B}) = 1_{A\otimes B}$ for all objects A and B.*

DEFINITION 2.5. *A* right dual *to an object X in a monoidal category $\mathcal{C}$ is an object X^* equipped with maps $\epsilon : X \otimes X^* \to I$ and $\eta : I \to X^* \otimes X$ such that the compositions*

$$X \overset{\rho^{-1}}{\to} X \otimes I \overset{X\otimes\eta}{\to} X \otimes (X^* \otimes X) \overset{\alpha^{-1}}{\to} (X \otimes X^*) \otimes X \overset{\epsilon\otimes X}{\to} I \otimes X \overset{\lambda}{\to} X$$

and

$$X^* \overset{\lambda^{-1}}{\to} I \otimes X^* \overset{\eta\otimes X^*}{\to} (X^* \otimes X) \otimes X^* \overset{\alpha}{\to} X^* \otimes (X \otimes X^*) \overset{X^*\otimes\epsilon}{\to} X^* \otimes I \overset{\rho}{\to} X^*$$

are both identity maps.

A left dual $^*X$ *is defined similarly, but with the objects placed on opposite sides of the monoidal product.*

This type of duality is an abstraction from the sort of duality which exists in categories of finite dimensional vector-spaces. It is not hard to show that the canonical isomorphism from the second dual of a vector-space to the space generalizes to give canonical isomorphisms $k : *(X^*) \to X$ and $\kappa : (^*X)^* \to X$. In general, however, there are not necessarily any maps from X^{**} or $^{**}X$ to X (cf. [**7**]). In cases where every object admits a right (resp. left) dual, it is easy to show that a choice of right (resp. left) dual for every object extends to a contravariant functor, whose application to maps will be denoted f^* (resp. $^*f$), and that the canonical maps noted above become natural isomorphisms between the compositions of these functors and the identity functor. Likewise, it is easy to show that $(A \otimes B)^*$ is canonically isomorphic to $B^* \otimes A^*$, and similarly for left duals.

In the case of a braided monoidal category every right dual is also a left dual, in general in a non-canonical way (cf. [**7**]). In symmetric monoidal categories, we return to the familiar: right duals are canonically left duals. For non-symmetric braided monoidal categories, the structure will be canonical only in the presence of additional structure. As we will be concerned only with the case of categories in which all objects admit duals, we make

DEFINITION 2.6. *A braided monoidal category $\mathcal{C}$ is* tortile (or ribbon) *if all objects admit right duals, and it is equipped with a natural isomorphism $\theta : 1_{\mathcal{C}} \Rightarrow 1_{\mathcal{C}}$, which satisfies*

$$\theta_{A\otimes B} = \sigma_{B,A}(\sigma_{A,B}(\theta_A \otimes \theta_B))$$

and

$$\theta_{A^*} = \theta_A^*$$

DEFINITION 2.7. *A symmetric monoidal category $\mathcal{C}$ is* rigid *if all objects admit (right) duals.*

For details of the canonicity of the correspondence between right and left duals in the tortile case see [**25**].

The importance of tortile categories for knot theory is provided by the extremely beautiful coherence theorem of Shum [**20**]:

THEOREM 2.8. *The tortile category freely generated by a single object X is monoidally equivalent to the category of framed tangles $\mathcal{FT}$.*

To be precise $\mathcal{FT}$ is the category whose objects are framed, oriented (i.e. signed) finite sets of points in $(0,1)^2$, and whose arrows are framed oriented compact 1-submanifolds of $[0,1]^3$ whose intersection with the boundary is normal and consists of finite sets of points lying in $(0,1)^2 \times \{0\}$ (the source) and $(0,1)^2 \times \{1\}$ (the target), modulo ambient isotopy rel boundary. Composition and $\otimes$ are defined by pasting on the third and second coordinates respectively and rescaling. Associativity and unit transformations are given by the obvious ascending submanifolds, while the braiding is given by ascending submanifolds which pass one in front of the other, and duality is given on objects by mirror-imaging in the second coordinate, and reversing all signs, with half curves-of-rotation as structure maps (cf. [**25**]).

This theorem gives rise to the following notion:

DEFINITION 2.9. *A* functorial invariant of framed links P_X *is an* $End(I)$*-valued invariant of framed links, where* I *is the unit object in a tortile category* $\mathcal{C}$*, obtained by choosing an object* X *in* $\mathcal{C}$*, and considering the restriction of the functor*

$$\mathcal{FT} \xrightarrow{G} \mathcal{F} \xrightarrow{F_X} \mathcal{C}$$

where $\mathcal{F}$ *is the free tortile category on one object generator,* G *is the equivalence functor from Shum's Coherence Theorem, and* F_X *is the functor induced by freeness which maps the generator to* X.

Numerous link invariants discovered since the mid 1980's are functorial invariants, including the Jones polynomial, and Zariski dense sets of values for the HOMFLY and Kauffman polynomials (the values "arising from quantum groups"), as is Conway's normalization of the classical Alexander polynomial.

The standard algebraic construction of these functorial invariants begins with the deformation of a universal enveloping algebra for a (simple) Lie algebra to produce a "quantum group", followed by the construction of the (tortile) category of its representations.

In [**25**] the author gave an exposition of a theorem of Deligne which dealt with direct deformation of a tensor category (starting with a Tannakian category). We will be concerned with the deformation theory of tensor categories and functors and its application to Vassiliev theory, so we review the necessary definitions:

DEFINITION 2.10. *An* infinitesimal deformation *of a* K*-linear tensor category* $\mathcal{C}$ *over an Artinian local* K*-algebra* R *is an* R*-linear tensor category* $\tilde{\mathcal{C}}$ *with the same objects as* $\mathcal{C}$*, but with* $Hom_{\tilde{\mathcal{C}}}(a,b) = Hom_{\mathcal{C}}(a,b) \otimes_K R$*, and composition extended by bilinearity, and for which the structure map(s)* α *(*ρ*, and* λ*, and if applicable* σ*,* η *and* ϵ*) reduce mod* $\mathfrak{m}$ *to the structure maps for* $\mathcal{C}$*, where* $\mathfrak{m}$ *is the maximal ideal of* R*. A deformation over* $K[\epsilon]/ < \epsilon^{n+1} >$ *is an* n^{th} order deformation.

Similarly an $\mathfrak{m}$-adic deformation *of* $\mathcal{C}$ *over an* $\mathfrak{m}$*-adically complete local* K*-algebra* R *is an* R*-linear tensor category* $\tilde{\mathcal{C}}$ *with the same objects as* $\mathcal{C}$*, but with* $Hom_{\tilde{\mathcal{C}}}(a,b) = Hom_{\mathcal{C}}(a,b)\widehat{\otimes}_K R$*, and composition extended by bilinearity and continuity, and for which the structure map(s)* α *(*ρ*, and* λ*, and if applicable* σ*,* η *and* ϵ*) reduce mod* $\mathfrak{m}$ *to the structure maps for* $\mathcal{C}$*, where* $\mathfrak{m}$ *is the maximal ideal of* R*. (Here* $\widehat{\otimes}_K$ *is the* $\mathfrak{m}$*-adic completion of the ordinary tensor product.) An* $\mathfrak{m}$*-adic deformation over* $K[[x]]$ *is a*formal series deformation.

Two deformations (in any of the above senses) are equivalent *if there exists a monoidal or semigroupal functor, whose underlying functor is the identity, and whose structure maps reduce mod* $\mathfrak{m}$ *to identity maps from one to the other.* The trivial deformation *of* $\mathcal{C}$ *is the deformation whose structure maps are those of* $\mathcal{C}$*. More generally, a deformation is* trivial *if it is equivalent to a trivial deformation.*

Finally, if $K = \mathbb{C}$ *(or* $\mathbb{R}$*), and all hom-spaces in* $\mathcal{C}$ *are finite dimensional, a* finite deformation *of* $\mathcal{C}$ *is a* K*-linear tensor category with the same objects and* *maps as $\mathcal{C}$, but with structure maps given by the structure maps of a formal series deformation evaluated at $x = \xi$ for some $\xi \in K$ such that the formal series defining all of the structure maps converge at ξ.*

Similarly, we can define deformations of monoidal and semigroupal functors targeted at tensor categories by

DEFINITION 2.11. *Given a K-linear tensor category $\mathcal{D}$. An* infinitesimal deformation *of a monoidal (resp. semigroupal) functor $F : \mathcal{C} \to \mathcal{D}$ over an Artinian local K-algebra R is a monoidal (resp. semigroupal) functor $F' : \mathcal{C} \to \tilde{\mathcal{D}}$ where the R-linear tensor category $\tilde{\mathcal{D}}$ is an infinitesimal deformation of $\mathcal{D}$ over R and for which the structure maps $\tilde{F}'$ and F'_0 (as well as those for $\mathcal{D}$) reduce mod $\mathfrak{m}$ to the structure maps for F (and $\mathcal{D}$), where $\mathfrak{m}$ is the maximal ideal of R. A deformation over $K[\epsilon]/ < \epsilon^{n+1} >$ is an nth* order deformation.

Similarly an $\mathfrak{m}$-adic deformation *of F over an $\mathfrak{m}$-adically complete local K-algebra R is a monoidal or semigroupal functor targeted at the R-linear tensor category $\tilde{\mathcal{D}}$ is an $\mathfrak{m}$-adic deformation of $\mathcal{D}$ and for which the structure maps reduce mod $\mathfrak{m}$ to the structure maps for F and $\mathcal{D}$, where $\mathfrak{m}$ is the maximal ideal of R. (Here $\widehat{\otimes_K}$ is the $\mathfrak{m}$-adic completion of the ordinary tensor product.) An $\mathfrak{m}$-adic deformation over $K[[x]]$ is a* formal series deformation.

Two deformations (in any of the above senses) are equivalent *if there exists a monoidal or semigroupal natural isomorphism between them which reduces mod $\mathfrak{m}$ to identity maps.* The trivial deformation *of F is the deformation whose structure maps are those of F. More generally, a deformation is* trivial *if it is equivalent to the trivial deformation.*

A deformation in any of the above senses will be called purely functorial *whenever the target category $\tilde{\mathcal{D}}$ is the trivial deformation of $\mathcal{D}$.*

In what follows, the category or functor we deform will generally have a more restrictive structure than its deformations. Thus, we will speak of a braided deformation or a tortile deformation of a rigid symmetric category or a semigroupal deformation of a monoidal functor, meaning one in which only the specified coherence conditions are satisfied.

Deligne's theorem [**25**] then says

THEOREM 2.12. *Any braided deformation of a Tannakian category $\mathcal{C}$ admits a unique tortile structure, and is thus a tortile deformation.*

In fact the proof given in [**25**] of this result will carry a more general result once it is observed that the use of intertwining matrices is unnecessary, and that it suffices to note that the map in question is the sum of the identity map with $\mathfrak{m}$-multiples of maps in $\mathcal{C}$.

THEOREM 2.13. *Any braided monoidal deformation of a rigid symmetric K-linear tensor category $\mathcal{C}$ for K any field (char $k \neq 2$) admits a unique tortile structure, and is thus a tortile deformation.*

These theorems are quite helpful, since they show that even for topological applications where the full tortile structure is required, it is sufficient to understand the structure of braided deformations.

We will also have cause to consider the notion of a reduction of a deformation.

DEFINITION 2.14. *If $\mathcal{C}$ is a K-linear semigroupal category, R an Artinian local ring (resp. an $\mathfrak{m}$-adically complete local ring), $\tilde{\mathcal{C}}$ a deformation ($\mathfrak{m}$-adic deformation) of $\mathcal{C}$ over R, and J an ideal of R, then the* reduction mod$J$ of $\tilde{\mathcal{C}}$, *denoted $\tilde{\mathcal{C}}/J$, is the category whose hom-spaces are $Hom_{\tilde{\mathcal{C}}}(X,Y) \otimes_R R/J$ with the obvious composition induced by that on $\tilde{\mathcal{C}}$, and structure maps obtained as the tensor product of the structure maps of $\tilde{\mathcal{C}}$ with $1 \in R/J$.*

DEFINITION 2.15. *If $\mathcal{C}$ is a K-linear semigroupal category, $F : \mathcal{D} \to \mathcal{C}$ a semigroupal functor, R an Artinian local ring (resp. an $\mathfrak{m}$-adically complete local ring), $F' : \mathcal{D} \to \tilde{\mathcal{C}}$ a deformation (resp. $\mathfrak{m}$-adic deformation) of F over R, and J an ideal of R, then the* reduction mod$J$ of F', *denoted F'/J, is the monoidal functor targetted at $\tilde{\mathcal{C}}/J$ and structure maps obtained as the tensor product of the structure maps of $\tilde{\mathcal{C}}$ with $1 \in R/J$.*

We will then need the following:

LEMMA 2.16. *If $\tilde{\mathcal{C}}$ is an nth order deformation of a K-linear semigroupal category $\mathcal{C}$ (resp. a formal series deformation) (perhaps with additional structure, e.g., monoidal, braided or tortile), then for $k < n$ (resp. for any $k \in \mathbb{N}$), $\tilde{\mathcal{C}}/ < \epsilon^{k+1} >$ is a kth order deformation of $\mathcal{C}$ with the same additional structure. Similarly if F' is an nth order deformation of a semigroupal functor targetted at $\mathcal{C}$ (resp. a formal series deformation), then for any $k < n$ (resp. any $k \in \mathbb{N}$) $F'/ < \epsilon^{k+1} >$ is a kth order deformation of F. If, moreover, F' is purely functorial, so is $F'/ < \epsilon^{k+1} >$.*

PROOF. It suffices to observe that the hom-spaces of the reduction are those of a kth order deformation, and that coherence conditions which hold before reduction mod J will still hold after reduction. □

Until recently the notion of a deformation of a tensor category was a definition without any attendant theorems, examples coming primarily from categories of representations of a deformed Hopf algebra (quantum group). In [**3**] an exposition was given of the deformation of tensor categories without braiding in terms of a cohomology theory for tensor categories. This theory is somewhat unsatisfactory in that it is unclear whether the obstructions are of a purely cohomological nature.

For our purposes, however, it is much more interesting to consider the deformation theory of tensor functors rather than tensor categories. On the one hand, the obstruction theory will prove to be purely cohomological, and on the other, it provides the correct notion of deformation for braidings thanks to the following results of Joyal and Street [**15**]:

DEFINITION 2.17. *A multiplication on a monoidal category $\mathcal{C}$ is a (strong) monoidal functor $(\Phi : \mathcal{C} \times \mathcal{C} \to \mathcal{C}, \tilde{\Phi}, \Phi_0)$ (usually denoted Φ by abuse of notation) together with monoidal natural isomorphisms $\mathfrak{r} : \Phi(Id_{\mathcal{C}}, I) \Rightarrow Id_{\mathcal{C}}$ and $\mathfrak{l} : \Phi(I, Id_{\mathcal{C}}) \Rightarrow Id_{\mathcal{C}}$*

THEOREM 2.18. *In a monoidal category $\mathcal{C}$, a family of arrows*

$$\sigma_{A,B} : A \otimes B \longrightarrow B \otimes A$$

is a braiding if and only if the following define a multiplication Φ on $\mathcal{C}$: $\Phi = \otimes$, $\Phi_0 = \rho_I^{-1}$, $\mathfrak{r} = \rho$, $\mathfrak{l} = \lambda$ and

$$\tilde{\Phi}_{(A,A'),(B,B')} = \lceil (1 \otimes \sigma) \otimes 1 \rceil : (A \otimes A') \otimes (B \otimes B') \longrightarrow (A \otimes B) \otimes (A' \otimes B')$$

Here, and in what follows, the notation $\lceil\ \rceil$ applied to a map in a monoidal category indicates pre- and post- composition by the unique map given by Mac Lane's coherence theorem [**17**] as a composite of "prolongations" of the structural transformations α, ρ, and λ so that the source and targets will be as specified. We will also apply this notation to sequences of maps, in which case, Mac Lane coherence maps may be inserted between each pair to provide a well-defined composition, and to both single maps and sequences of maps in settings where a monoidal functor is involved. In this latter setting the coherence map(s) may also involve structural transformations from the monoidal functor (see Epstein [**5**] for the theorems which justify use of this notation in the case of monoidal functors).

A converse theorem also due to Joyal and Street states

THEOREM 2.19. *For any multiplication Φ on a monoidal category $\mathcal{C}$, a braiding σ for $\mathcal{C}$ is defined by*

$$\sigma_{A,B} = \mathfrak{l} \otimes \mathfrak{r}(\tilde{\Phi}^{-1}(\Phi(\rho^{-1}, \lambda^{-1})(\Phi(\lambda, \rho)(\tilde{\Phi}(\mathfrak{l}^{-1} \otimes \mathfrak{r}^{-1})))))$$

The multiplication obtained from σ via the previous theorem is isomorphic (an the obvious sense) to Φ, and if τ is any braiding on $\mathcal{C}$, and Φ is the multiplication obtained from τ via the previous theorem, then $\sigma = \tau$.

As noted above, the importance of these theorems for us lies in the fact that they reduce the question of studying braidings to the study of monoidal functors, and it is from this point of view that we will examine the deformation theory of braided tensor categories.

3. The Deformation Theory of Semigroupal and Monoidal Functors and Braided Monoidal Categories

We now turn our attention to the actual structure of deformations of braided monoidal categories. As we have stated above, we proceed by the indirect route of developing a deformation theory for monoidal functors, then applying it to classify deformations of a multiplication on a tensor category.

As in [**3**] our starting point is the observation that the coherence condition for the structure at hand is formally a cohomological condition written multiplicatively. As was the case in [**3**] for the structure maps of a deformation of a tensor category, if we consider the structure maps for a purely functorial first order deformation of a semigroupal functor, we find that the main coherence condition is equivalent (in the presence of the coherence for the map being deformed) to a familiar cohomological condition, or rather, to what would be a familiar condition were it not for the intervening $\lceil\ \rceil$'s.

In the present case, if $\tilde{F}' = \tilde{F} + F^{(1)}\epsilon$ for $F^{(1)}$ a natural transformation from $F(-\otimes -)$ to $F(-) \otimes F(-)$ and $\epsilon^2 = 0$, then the hexagon on $\tilde{F}'$ is equivalent to the condition that

$$\lceil F^{(1)}_{A\otimes B,C} \rceil + \lceil F^{(1)}_{A,B} \otimes F(C) \rceil = \lceil F^{(1)}_{A,B\otimes C} \rceil + \lceil F(A) \otimes F^{(1)}_{B,C} \rceil$$

recalling that here $\lceil\ \ \rceil$ denotes "padding" with coherence maps for F and the semigroupal structures, as is permitted by the results of [**5**]. Of course, solving for 0 gives us the condition that $\delta(F^{(1)}) = 0$, where δ is the familiar coboundary for the bar resolution, but dressed up with $\lceil\ \rceil$'s on its terms, and using $\otimes$ in place of multiplication.

We now wish to describe a setting in which this "coboundary" is in fact the coboundary in a cochain complex associated to the semigroupal functor F. As in [**3**] let ${}^n\otimes$ (resp. $\otimes^n$) denote the completely left (resp. right) parenthesized n-fold iterated semigroupal product. It should then be observed that all of the terms of the formula above are natural transformations from $F({}^3\otimes)$ to $\otimes^3(F^3)$. It is thus reasonable to make

DEFINITION 3.1. *The* deformation complex *of a semigroupal functor targetted at a K-linear tensor category $\mathcal{C}$ is the K-cochain complex $(X^\bullet(F), \delta)$ for which*

$$X^n(F) = Nat(F({}^n\otimes), \otimes^n(F^n))$$

with coboundary δ given by the usual bar-resolution formula with $\otimes$ as multiplication, and $\lceil\quad\rceil$ applied to all terms. Denote the cohomology of the complex by $H^\bullet(F)$

It follows from the usual argument (and the bilinearity of the composition of natural transformations and $\otimes$ and the coherence conditions for F and $\otimes$) that $\delta^2 = 0$. It is also easy to see that a semigroupal natural isomorphism which reduces to the identity mod ϵ between two first order deformations is precisely a 1-cochain which cobounds the difference between the two 2-cocycles which name the deformations. Thus we have

THEOREM 3.2. *There is a natural 1-1 correspondence between the first order purely functorial deformations of a semigroupal functor F and the 2-cocycles of the deformation complex of F. Moreover, the semigroupal natural isomorphism classes of first order purely functorial deformations of F are in natural 1-1 correspondence with $H^2(F)$.*

Thus, it is precisely when $H^2(F) \neq 0$ that F admits non-trivial purely functorial deformations. Because this rigidity is limited to purely functorial deformations, we will say a semigroupal functor F is *purely rigid* whenever $H^2(F) = 0$.

We can now consider the question of extending an nth order deformation to an $(n+1)^{st}$ order deformation, that is, suppose we have an nth order deformation

$$F_n = F + F^{(1)}\epsilon + \ldots + F^{(n)}\epsilon^n$$

where $\epsilon^{n+1} = 0$, and we wish to find an $n+1^{st}$ order deformation

$$F_n = F + F^{(1)}\epsilon + \ldots + F^{(n+1)}\epsilon^{n+1}$$

(for $\epsilon^{n+2} = 0 \neq \epsilon^{n+1}$) by adding another term.

It is easy to see that the required condition on $F^{(n+1)}$ is that

$$\begin{aligned}\delta(F^{(n+1)}) = &\sum_{l+m=n+1, 1\leq l,m\leq n} \lceil F^{(m)}_{A,B} \otimes F(C)(F^{(l)}_{A\otimes B,C})\rceil \\ &- \sum_{p+q=n+1, 1\leq p,q\leq n} \lceil F(A) \otimes F^{(q)}_{B,C}(F^{(p)}_{A,B\otimes C})\rceil\end{aligned}$$

Thus this right-hand side may be viewed as the obstruction to extending the nth order deformation to an $(n+1)^{st}$ order deformation. As in [**9**], we will see that this obstruction is always a 3-cocycle, and thus the obstruction to extending a purely functorial deformation by one order may be viewed as a cohomology class in $H^3(F)$.

In fact, the cochain complex associated to a semigroupal functor shares many of the properties of the Hochschild complex of an associative algebra A with coefficients in A which were described by Gerstenhaber [**8, 9, 10**]. In particular, we have two products defined on cochains:

$$- \cup - : X^n(F) \times X^m(F) \to X^{n+m}(F)$$

given by

$$G \cup H_{A_1,\dots A_{n+m}} = \lceil G_{A_1,\dots,A_n} \otimes H_{A_{n+1},\dots,A_{n+m}} \rceil$$

and

$$\langle -, - \rangle : X^n(F) \times X^m(F) \to X^{n+m-1}(F)$$

given by

$$\langle G, H \rangle_{A_1,\dots A_{n+m-1}} = \sum (-1)^{mi} \lceil F(A_1) \otimes \dots F(A_i) \otimes H_{A_{i+1},\dots,A_{i+n}} \otimes F(A_{i+n+1}) \otimes \dots \otimes F(A_{n+m-1}) (G_{A_1,\dots,A_i,A_{i+1}\otimes\dots\otimes A_{i+n},A_{i+n+1},\dots A_{n+m-1}}) \rceil$$

PROPOSITION 3.3. *This latter product comes from a "pre-Lie system" in the terminology of Gerstenhaber* [**8**] *given by*

$$\langle G, H \rangle^{(i)}_{A_1,\dots A_{n+m-1}} = \lceil F(A_1) \otimes \dots F(A_i) \otimes H_{A_{i+1},\dots,A_{i+n}} \otimes F(A_{i+n+1}) \otimes \dots \otimes F(A_{n+m-1}) (G_{A_1,\dots,A_i,A_{i+1}\otimes\dots\otimes A_{i+n},A_{i+n+1},\dots A_{n+m-1}}) \rceil$$

where $X^n(F)$ has degree $n-1$.

PROOF. First, note that the ambiguities of parenthesization in the semigroupal products in this definition are rendered irrelevant by the $\lceil\ \rceil$ on each term by virtue of the coherence theorem for semigroupal functors.

It is obvious that the product is given by a sum of these terms, so the content of the proposition is really that the $\langle -, - \rangle^{(i)}$'s satisfy the definition of a pre-Lie system, that is for $G \in X^m(F)$, $H \in X^n(F)$, and $K \in X^p(F)$ that

$$\langle \langle G, H \rangle^{(i)}, K \rangle^{(j)} = \begin{cases} \langle \langle G, K \rangle^{(j)}, H \rangle^{(i+p-1)} & \text{if } 0 \leq j \leq i-1 \\ \langle G, \langle H, K \rangle^{(j-i)} \rangle^{(i)} & \text{if } i \leq j \leq n \end{cases}$$

(Recall that a k-chain has degree $k-1$.)

This is a simple computational check. One must remember that naturality will allow one to commute the prolongations of K and H in verifying the first case. □

We can then write the obstruction equation as

$$\delta(F^{(n+1)}) = \sum_{i=1}^{n} \langle F^{(i)}, F^{n-i+1} \rangle$$

In particular, we have a

PROPOSITION 3.4. *For any n^{th} order purely functorial deformation of a semigroupal functor F targeted at a K-linear tensor category, the obstruction to extension to an $(n+1)^{st}$ order deformation is a 3-cocycle.*

PROOF. This follows immediately from the proof in Gerstenhaber [**9**] of §5, Proposition 3, since that proof depends only on the fact that the operation comes from a pre-Lie system. □

In particular, if $H^3(F) = 0$, any first order purely functorial deformation of F can be extended to an nth order deformation for any n, and thus to a formal series deformation.

The case of a monoidal functor is somewhat less clean than for semigroupal functors. Here we will only consider the case of those purely functorial deformations for which the unit isomorphism F_0 is deformed trivially, that is, is given by $F_0 \otimes 1$, where 1 is the unit of the ring R over which the deformation takes place. We call such deformations *proper.*

It appears to be possible to remove this restriction in the case of monoidal functors which are faithful and monic on objects (or equivalent to such a functor), but we will not pursue this further in this paper.

A brief consideration of the unit coherence conditions for monoidal functors shows that a proper n^{th} order deformation of a monoidal functor must be given by a semigroupal deformation

$$\tilde{F}' = \tilde{F}' + F^{(1)}\epsilon + \ldots + F^{(n)}\epsilon^n$$

for which $F^{(k)}_{A,I} = 0$ and $F^{(k)}_{I,A} = 0$ for all k and all objects A.

This leads to the following

DEFINITION 3.5. *The* proper deformation complex *of a monoidal functor targetted at a K-linear tensor category is the K-cochain complex $(C^\bullet(F), \delta)$ for which*

$$C^n(F) = \{\phi_{A_1,\ldots,A_n} | A_i = I \vdash \phi_{A_1,\ldots,A_n} = 0\} \subset Nat(F(^n\otimes), \otimes^n(F^n))$$

with coboundary δ given by the restriction of the coboundary on the deformation complex of F. We denote the cohomology of this complex by $\mathbf{H}^\bullet(F)$.

The following proposition ensures that this definition makes sense and that the proper deformation complex has the same type of structures as the deformation complex:

PROPOSITION 3.6. *If G and H are proper cochains, then so are $\delta(G)$, $G \cup H$, and $\langle G, H\rangle^{(i)}$.*

PROOF. For the latter two, observe that any I will give a factor in the expression for the product in which I occurs either as an index of G or as an index of H, and thus the product will be 0. For the first, note that all terms of $\delta(G)$ except two will be of the form $\lceil G \rceil$ for an instance of G with I as one of the indices. The remaining two terms will be of the same form (though without I indices), of opposite signs, and will differ only in which indices are tensored by I. Thus by the presence of the $\lceil\ \rceil$ they will be the same and will cancel. □

Notice in particular that the obstructions to extending a proper deformation will always be proper cochains.

It then follows immediately that we have results corresponding to Theorem 3.2 and Proposition 3.4.

THEOREM 3.7. *There is a natural 1-1 correspondence between the first order proper deformations of a monoidal functor F and the 2-cocycles of the proper deformation complex of F. Moreover, the monoidal natural isomorphism classes of first order proper deformations of F are in natural 1-1 correspondence with $\mathbf{H}^2(F)$.*

PROPOSITION 3.8. *For any n^{th} order proper deformation of a monoidal functor F targetted at a K-linear tensor category, the obstruction to extension to an $n+1^{st}$ order deformation is a 3-cocycle.*

We may now speak of a monoidal functor as *properly rigid* if $\mathbf{H}^2(F) = 0$, and observe that if $\mathbf{H}^3(F) = 0$ then any first order proper deformation can be extended to a formal series deformation.

Now, consider the case of a multiplication on a tensor category $\mathcal{C}$. The requirement that there exist right and left unit transformations for the multiplication imposes an additional condition on the admissible deformations. However, these also are readily understood in terms of the cohomology theory.

In particular, notice that if F and G are a composable pair of monoidal (or semigroupal) functors, a deformation of either leads to an induced deformation of the composite (provided the appropriate target categories are K-linear tensor categories). More precisely, if $\tilde{F}$ is a deformation of F, then $\tilde{F}(G)$ is a deformation of $F(G)$, while if $\tilde{G}$ is a deformation of G, and both the source and target of F are K-linear tensor categories, then $\tilde{F}_{\text{trivial}}(\tilde{G}_!)$ is a deformation of $F(G)$, where $\tilde{F}_{\text{trivial}}$ denotes the trivial deformation of F and $\tilde{G}_!$ denotes the R-linear extension of $\tilde{G} : \mathcal{C} \to \tilde{\mathcal{D}}$ to $\tilde{\mathcal{C}}$. In terms of induced deformations we then have

THEOREM 3.9. *If Φ is a multiplication on a tensor category $\mathcal{C}$, and Φ' is a proper deformation of Φ as a monoidal functor then Φ' is a multiplication on $\tilde{\mathcal{C}}$ precisely when the induced deformations $\Phi'(-, I)$ and $\Phi'(I, -)$ are trivial.*

PROOF. If the induced deformations are trivial, we can compose the trivializing natural isomorphism with the R-bilinear extension of $\mathfrak{r}$ (resp. $\mathfrak{l}$) to obtain the deformed $\mathfrak{r}$ (resp. $\mathfrak{l}$). Conversely, if there are deformed $\mathfrak{r}$ and $\mathfrak{l}$, their composition with the R-bilinear extension of the inverse to the undeformed $\mathfrak{r}$ and $\mathfrak{l}$ give the respective trivializations. □

4. Vassiliev Invariants for Framed Links

It is usual to discuss Vassiliev theory in terms of unframed oriented knots and links (cf. [**2**], [**21**], [**23**]). We will, however, remain in the setting most natural for functorial invariants (and incidentally most closely connected to 3- and 4-manifold topology), that of framed links.

Following Goryunov [**11**] one can regard framed knots (and links) not as ordinary knots (and links) with additional structure, but as equivalence classes of mappings from open annular neighborhoods U of S^1 (or disjoint unions of such) into $\mathbb{R}^3$. Observe that any such mapping $g : U \to \mathbb{R}^3$ induces a mapping Tg from $i^*(T\mathbb{R}^2)$ to $T\mathbb{R}^3$. (Here i is the inclusion of S^1 or a disjoint union of S^1's into U.)

For ease in the link setting, we need to specify a countable family of disjoint circles: say $S_n = \{z \mid |z - 3n| = 1\} \subset \mathbb{C}$, then consider disjoint annular neighborhoods of $S_1 \cup S_2 \cup \cdots \cup S_k$ when dealing with k-component links.

DEFINITION 4.1. *Two mappings $g_i : U_i \to \mathbb{R}^3$ $i = 1, 2$ are* equivalent *if the U_i's are annular neighborhoods of the same $S_1 \cup \cdots \cup S_k$ and the mappings $Tg_i : i^*(T\mathbb{R}^2) \to T\mathbb{R}^3$ coincide.*

The space of (possibly singular) framed links is the space of all C^∞ mappings of U's modulo this equivalence relation. We denote it by Ω_f. The subspace of equivalence classes of mapping of neighborhoods of $S_1 \cup \cdots \cup S_k$ is the space of (possibly singular) framed links of k components, and will be denoted $\Omega_f(k)$. Now consider the subspace $\mathcal{O}_f$ of all (equivalence classes of) mappings such that Tg is an embedding.

DEFINITION 4.2. *A* (non-singular) framed link *is a connected component of* $\mathcal{O}_f$. *A connected component of* $\mathcal{O}_f(k) = \mathcal{O}_f \cap \Omega_f(k)$ *is a* (non-singular) framed link of k components.

First, note that we will often drop the adjective "non-singular" to match the usual usage in knot theory (as in earlier sections). Second, the designation of these maps as non-singular implicitly identifies the element of $\Omega_f \setminus \mathcal{O}_f$ as *singular*. We denote the discriminant locus $\Omega_f \setminus \mathcal{O}_f$ by Σ_f. As observed in Goryunov [**11**] the discriminant is the union of two hypersurfaces, one on which the framing degenerates, denoted Σ'_f, and one on which the disjoint union of circles is not embedded, denoted Σ''_f.

The minimal degenerations of each type are illustrated in Figure 1. In the top line of Figure 1 the diagonal line indicates the framing curve (the edge of the ribbon obtained by applying the exponential map for the standard metric to the framing vectors), while in the bottom line we suppress indication of the framing. We denote the subspace of Σ_f in which there are exactly n degenerations of either type by $\Sigma_{f,n}$, and its intersection with $\Omega_f(k)$ by $\Sigma_{f,n}(k)$.

As in Goryunov [**11**], we coörient the finite codimensional strata of Σ_f by the local prescriptions given in Figure 1, and give a Vassiliev type prescription for the extension of invariants of framed knots to singular framed knots as in Figure 2.

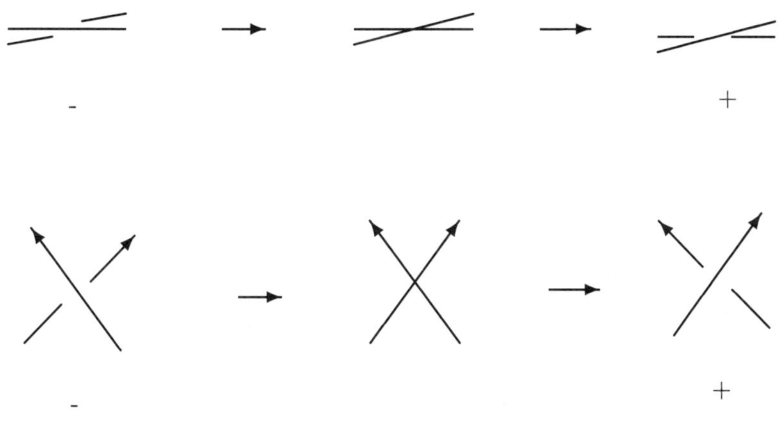

FIGURE 1. Coörienting the finite codimensional strata

DEFINITION 4.3. *A Vassiliev invariant of links (resp. links of k components) is a locally constant function on $\mathcal{O}_f$ (resp. $\mathcal{O}_f(k)$) whose extension according to the prescription of Figure 2 vanishes on $\Sigma_{f,n+1}$ (resp. $\Sigma_{f,n+1}(k)$) for some n, in which case the invariant is said to be of type $\leq n$. The invariant is of type N when N is the minimal such n.*

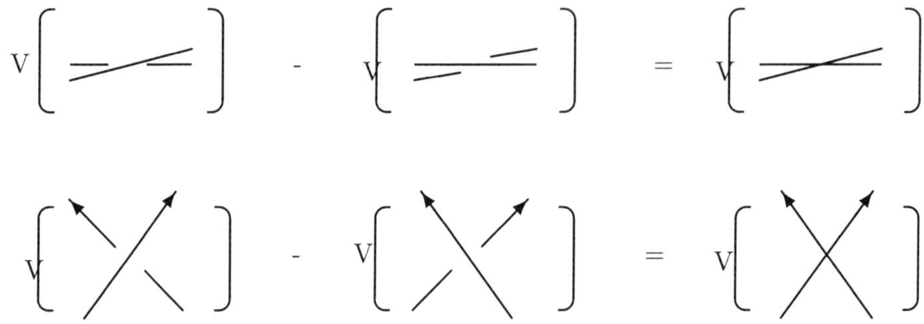

FIGURE 2. Vassiliev type extension formulae

5. Deformations and Vassiliev Invariants

We are now in a position to state our main theorem:

THEOREM 5.1. *Let $\mathcal{C}$ be any K-linear rigid symmetric monoidal category, and let $\tilde{\mathcal{C}}$ be any n^{th} order tortile deformation of $\mathcal{C}$. For any object X of $\tilde{\mathcal{C}}$, let V_X denote the functor from $\mathcal{FT}$ to $\tilde{\mathcal{C}}$ induced by Shum's Coherence Theorem. Then V_X restricted to $End(I)$, regarded as the set of framed links, is a $K[\epsilon]/ < \epsilon^{n+1} >$-valued Vassiliev invariant of type $\leq n$, and is, moreover, multiplicative under disjoint union.*

From this will follow, almost as a corollary

THEOREM 5.2. *Let $\mathcal{C}$ be any K-linear rigid symmetric monoidal category, and let $\tilde{\mathcal{C}}$ be any n^{th} order tortile deformation of $\mathcal{C}$ (resp. formal series deformation of $\mathcal{C}$). For any object X of $\tilde{\mathcal{C}}$, let V_X denote the functor from $\mathcal{FT}$ to $\tilde{\mathcal{C}}$ induced by Shum's Coherence Theorem, and let $V_{X,k}$ denote the K-valued framed link invariant which assigns to any link the coefficient of ϵ^k, for $k = 0, \ldots, n$, (resp. for $k \in \mathbb{N}$). Then $V_{X,k}$ is a Vassiliev invariant of type $\leq k$.*

The proof of Theorem 5.1 is quite simple and similar to previous proofs of similar results:

The key is to observe that the bilinearity of composition in $\tilde{\mathcal{C}}$ allows us to use the Vassiliev prescription to extend the functor V_X from $\mathcal{FT}$ to a larger category of singular framed tangles, $\widetilde{\mathcal{FT}}$, whose maps are isotopy classes of framed tangles with finitely many degeneracies of either of the two basic types.

Consider a singular framed link with $n+1$ degeneracies (of either type). Now, we can represent the framed link as a composition of singular framed tangles, each of which has at most one degeneracy, crossing, framing twist, or extremum. Now the value of the extended functor on such a tangle with a degeneracy of the first type (framing degeneracy) is a monoidal product of identity maps with $\theta_X - \theta_X^{-1}$ (or its dual) , while the value on such a tangle with a degeneracy of the second type is a monoidal product of identity maps with $\sigma_{X,X} - \sigma_{X,X}^{-1}$.

Now, observe that

$$\theta_X - \theta_X^{-1} \in Hom_{\mathcal{C}}(X,X) \otimes < \epsilon >$$

and

$$\sigma_{X,X} - \sigma^{-1}_{X,X} \in Hom_{\mathcal{C}}(X \otimes X, X \otimes X) \otimes < \epsilon > .$$

It follows from the bilinearity of composition in $\tilde{\mathcal{C}}$ that the composite representing the framed link as an element of $End(I)$ lies in $End_{\mathcal{C}} \otimes < \epsilon^{n+1} >= 0$, thus showing V_X to be Vassiliev of type $\leq n$. Multiplicativity follows from functoriality.

Theorem 5.2 follows from the Theorem 5.1, Lemma 2.16, and the following lemma, the proof of which is a trivial exercise:

LEMMA 5.3. *If V is an A-valued Vassiliev invariant, and $f : A \rightarrow B$ is a linear map (A and B here are abelian groups) then $f(V)$ is a B-valued Vassiliev invariant.*

On general principles these results are more general than previous results of this form: our initial category can be any K-linear rigid symmetric monoidal category for any field K, not just the category of representations of a (semi)simple Lie algebra. In particular, this means that the invariants obtained from any representation of a q-deformed universal enveloping algebra for a super-Lie algebra (cf. [**18**]) give rise to sequences of Vassiliev invariants for links of any number of components when we set $q = e^h$ (or any other formal series for that matter) and consider the coefficients of h^n. (The observation about using other series, of course applies in the ordinary Lie algebra case as well.) Likewise the coefficients of any h^n in invariants arising from a representation of a multiparameter quantum group (cf. [**4**], [**13**], [**22**])and *any* instantiation of the deformation parameters by power series will be Vassiliev invariants (even in the case of quasi-Hopf deformations). In fact, by repeating the arguments given above for quotients of $K[[x]]$ by nilpotent ideals in the case of $K[[h_1, \ldots, h_n]]$, one can show that the coefficients of any monomial in $h_1, \ldots, h_n$ in the invariant associated to a representation of a multiparameter quantum group and any instantiation of the ith deformation parameter by a power series in h_i are Vassiliev invariants.

6. Questions Raised

The results contained herein are only partial. In order to fully relate Vassiliev theory to the deformation theory of braided monoidal categories, it will be necessary to give a good accounting of that deformation theory in the non-proper case. The difficulties here are two-fold. First is the removal of the restriction that the unit isomorphism not be deformed, and second is the removal of the purely functorial restriction.

We thus ask:

1. Is there a good cohomological deformation theory for purely functorial (but not necessarily proper) deformations of monoidal functors?
2. Can the deformation theory of this work be combined in a satisfactory way with the deformation theory of monoidal categories given in [**3**] to provide a deformation theory for (not necessarily purely functorial) deformations of monoidal functors?
3. Even a satisfactory answer to question 2. will not suffice to give a complete deformation theory for braided monoidal categories, since in 2. only the target is deformed, while a deformation of the structure maps of the underlying category, induce deformations of the structure maps of both the

source and the target of the multiplication. Thus, we ask: Is there a satisfactory cohomological deformation theory for unrestricted deformations of multiplications on monoidal categories?

Likewise, the construction given here of Vassiliev invariants from deformations of monoidal categories leads to a series of questions regarding the converse problem: which Vassiliev invariants arise from functorial invariants?

Chief among these are various precise formulations of that basic question:

1. Given a K-valued Vassiliev invariant v_n of type n of framed knots, when do there exist Vassiliev invariants $v_0, \dots, v_{n-1}$ of types $0, \dots, n-1$ respectively, such that the $K[\eta]/ < \eta^n + 1 >$-valued invariant
$$v_0 + v_1\eta + \dots + v_n\eta^n$$
is the restriction to framed knots of an invariant defined by an object in some tortile deformation of a K-linear rigid symmetric monoidal category?
2. Given a K-valued Vassiliev invariant v of type n of m-component links,
   (a) When does v extend to a Vassiliev invariant of links as defined above?
   (b) When do there exist Vassiliev invariants of lower orders so that the generating function as in question 1. is the invariant defined by an object in some tortile deformation of a K-linear rigid symmetric monoidal category?
3. Given a Vassiliev invariant of links as described above, when can it be completed as in questions 1. and 2. by adding lower order Vassiliev invariants to give a functorial invariant?

A partial result toward answers to these three questions can be read off from the results above, namely, a necessary condition for a Vassiliev invariant of framed links to arise as a series coefficient in a tortile deformation of a symmetric tensor category:

THEOREM 6.1. *If v is a Vassiliev invariant of (framed) links arising as the coefficient of ϵ^n in an nth order tortile deformation of a rigid symmetric tensor category, then there exist Vassiliev invariants v_k of (framed) links $k = 1, \dots, n$ (with $v_n = v$) such that*
$$v(K \coprod L) = v_0(K)v_n(L) + v_1(K)v_{n-1}(L) + \dots v_n(K)v_0(L)$$
and each v_i is of type $\leq i+1$, where $\coprod$ denotes separated union of links.

In terms of Vassiliev invariants for links of a fixed number of components, we can similarly see

THEOREM 6.2. *If v is a Vassiliev invariant of (framed) links of N-components arising as the coefficient of ϵ^n in a tortile deformation of a rigid symmetric tensor category, then for all $M = 1, \dots, N$ and $i = 0, \dots, n$ there exist Vassiliev invariants $v_{M,i}$ of M component (framed) links of type $\leq i$ such that for all $K \coprod L$ separated unions of links, with K having M components and L having $N - M$ components, we have*
$$v(K \coprod L) =$$
$$v_{M,0}(K)v_{N-M,n}(L) + v_{M,1}(K)v_{N-M,n-1}(L) + \dots + v_{M,n}(K)v_{N-M,0}(L)$$

Finally, there are questions of deeper relationships between the cohomological deformation theories of of monoidal categories and of (bi)algebras and Vassiliev theory:

1. What is the relationship between the deformation theory of braided monoidal categories and the original cohomological formulation of Vassiliev invariants [**23**]?
2. What relationship is there between the Vassiliev invariants arising from the tortile deformations of a particular symmetric monoidal category and the structure of the category itself?
3. What is the relationship between the deformation theory of a (bi)algebra A and that of monoidal functors associated to the algebra (in particular, the forgetful functor

$$U : Rep(A) \rightarrow K - vect,$$

and for (quasi-)triangular A the tensor product as a monoidal functor with stuctural transformations as induced by the braiding or symmetry as in Theorem 2.18?

References

[1] D. Altschuler, *Representations of knot groups and Vassiliev invariants,* J. Knot Theory Ramifications **5** (1996), 421–426.

[2] J. Birman and X.-S. Lin, *Knot polynomials and Vassiliev's invariant,* Invent. Math. **111** (1993), 225–270.

[3] L. Crane and D. N. Yetter, *Deformations of (bi)tensor categories,* Cahiers Topologie Géom. Differentielle Catég., to appear.

[3a] A. A. Davydov, *Twisting of monoidal structures,* preprint, 1997.

[4] R. Dipper and S. Donkin, *Quantum GL_n,* Proc. London Math. Soc. **63** (1991), 162–211.

[5] D. B. A. Epstein, *Functors between tensored categories,* Invent. Math. **1** (1966), 221–228.

[6] P. J. Freyd and D.N. Yetter, *Braided compact closed categories with applications to low dimensional topology,* Adv. Math. **77** (1989), 156–182.

[7] P. J. Freyd and D. N. Yetter, *Coherence theorems via knot theory,* J. Pure Appl. Algebra **78** (1992), 49–76.

[8] M. Gerstenhaber, *The cohomology of an associative ring,* Ann. Math. **78** (1963), 267–288.

[9] M. Gerstenhaber, *On the deformation of rings and algebras,* Ann. Math. **79** (1964), 59–103.

[10] M. Gerstenhaber and D. Schack, *Algebraic cohomology and deformation theory,* in "Deformation theory of algebras and structure and applications (Il Ciocco, 1986)," NATO Adv. Sci. Inst. Ser. C: Math. Phys. Sci., 247, Kluwer Acad. Publ., Dordrecht, 1988, pp. 11–264.

[11] V. Goryunov, *Vassiliev invariants of knots in $\mathbb{R}^3$ and in a solid torus,* preprint, 1995.

[12] P. J. Freyd, D. N. Yetter, W. B. R. Lickorish, K. C. Millet, J. Hoste, and A. Ocneanu, *A new polynomial invariant of knots and links,* Bull. Amer. Math. Soc. **12** (1985), 239–246.

[13] M. Hazewinkel, *Multiparameter quantum groups and multiparameter R-matrices,* Acta Appl. Math. **41** (1995), 57–98.

[14] V. F. R. Jones, *A polynomial invariant of knots via vonNeumann algebras,* Bull. Amer. Math. Soc. **12** (1985), 103–111.

[15] A. Joyal and R. Street, *Braided tensor categories,* Adv. Math. **102** (1993), 20–78.

[16] L. H. Kauffman, *An invariant of regular isotopy,* Trans. Amer. Math. Soc. **318** (1990), 317–371.

[17] Mac Lane, S., "Categories for the Working Mathematician," Springer Verlag, New York-Berlin, 1971.

[18] Yu. I. Manin, *Multiparameter quantum deformations of the general linear supergroup,* Commun. Math. Phys. **123** (1989), 163–175.

[19] S. Piunikhin, *Weights of Feynman diagrams, link polynomials, and Vassiliev knot invariants,* J. Knot Theory Ramifications **4** (1995), 163–188.

[20] M.-C. Shum, *Tortile tensor categories,* doctoral dissertation, Macquarie University, 1989.

[21] T. Stanford, *Finite-type invariants of knots, links, and graphs,* Topology **35** (1996), 1027–1050.

[22] M. Takeuchi, *A two parameter quantization of GL_n,* Proc. Japan Acad. Ser. A **66** (1990), 112–114.

[23] V. A. Vassiliev, *Cohomology of knot spaces,* in "Theory of singularities and its applications," Adv. Sov. Math., 1, Amer. Math. Soc., Providence, RI, 1990, pp. 23–69.

[24] D. N. Yetter, *Topological quantum field theories associated to finite groups and crossed G-sets,* J. Knot Theory Ramifications **1** (1992), 1–20 .

[25] D. N. Yetter, *Framed tangles and a theorem of Deligne on braided deformations of Tannakian categories,* in "Deformation theory and quantum groups with applications to mathematical physics (Amherst, MA, 1990)," Contemp. Math., 134, Amer. Math. Soc., Providence, RI, 1992, pp. 325–349.

Department of Mathematics, Kansas State University, Manhattan, KS 66506
E-mail address: dyetter@math.ksu.edu